TC 18-06

Special Forces Guide to Information Operations

March 2013

DISTRIBUTION RESTRICTION: Distribution authorized to U.S. Government agencies and their contractors only to protect technical or operational information from automatic dissemination under the International Exchange Program or by other means. This determination was made on 30 March 2012. Other requests for this document must be referred to Commander, United States Army John F. Kennedy Special Warfare Center and School, ATTN: AOJK-CDI-SF, 3004 Ardennes Street, Stop A, Fort Bragg, NC 28310-9610.

DESTRUCTION NOTICE: Destroy by any method that will prevent disclosure of contents or reconstruction of the document.

FOREIGN DISCLOSURE RESTRICTION (FD 6): This publication has been reviewed by the product developers in coordination with the United States Army John F. Kennedy Special Warfare Center and School foreign disclosure authority. This product is releasable to students from foreign countries on a case-by-case basis only.

Headquarters, Department of the Army

This publication is available at Army Knowledge Online (htttps://armypubs.us.army.mil/doctrine/index.html). To receive publishing updates, please subscribe at http://www.apd.army.mil/AdminPubs/new_subscribe.asp.

TC 18-06

Training Circular
No. 18-06

Headquarters
Department of the Army
Washington, DC, 22 March 2013

Special Forces Guide to Information Operations

Contents

		Page
	PREFACE	vi
Chapter 1	FUNDAMENTALS	1-1
	Information Operations	1-2
	The Information Environment	1-3
	Information Superiority	1-4
	Information Operations Capabilities	1-5
	Considerations	1-5
	Conclusion	1-5
Chapter 2	INFORMATION OPERATIONS CAPABILITIES AND TACTICS	2-1
	Operations Security	2-2
	Military Deception	2-5
	Military Information Support Operations	2-11
	Electronic Warfare	2-15
	Computer Network Operations	2-16
	Combat Camera	2-17
	Local Populace and Key-Leader Engagements	2-19
	Countering Adversary Information Activities	2-26
	Rewards Programs	2-29
	Civil-Military Operations	2-30
	Public Affairs	2-31
	Defense Support to Public Diplomacy	2-31

Distribution Restriction: Distribution authorized to U.S. Government agencies and their contractors only to protect technical or operational information from automatic dissemination under the International Exchange Program or by other means. This determination was made on 30 March 2012. Other requests for this document must be referred to Commander, United States Army John F. Kennedy Special Warfare Center and School, ATTN: AOJK-CDI-SF, 3004 Ardennes Street, Stop A, Fort Bragg, NC 28310-9610.

Destruction Notice: Destroy by any method that will prevent disclosure of contents or reconstruction of the document.

Foreign Disclosure Restriction (FD 6): This publication has been reviewed by the product developers in coordination with the United States Army John F. Kennedy Special Warfare Center and School foreign disclosure authority. This product is releasable to students from foreign countries on a case-by-case basis only.

Contents

Chapter 3	PLANNING INFORMATION OPERATIONS	3-1
	The Staff Estimate for Information Operations	3-2
	Planning Considerations During Mission Analysis	3-2
	Mission Analysis Work Sheet	3-7
	Course of Action Development	3-7
	Considerations	3-19
	Consequence Management	3-19
Chapter 4	EXECUTION OF INFORMATION OPERATIONS	4-1
	Monitoring	4-1
	Evaluating	4-1
	Adjusting	4-3
	Reporting	4-8
Chapter 5	INTELLIGENCE SUPPORT TO INFORMATION OPERATIONS	5-1
	Information Operations and the Intelligence Cycle	5-2
	Intelligence "Push" and "Pull"	5-2
	Requests for Information	5-3
	Intelligence Preparation of the Operational Environment	5-3
	Visualizing the Information Environment	5-3
	The Combined Information Overlay	5-5
	Adversary Operations in the Information Environment	5-6
	Templating Using Center-of-Gravity Analysis	5-7
	Adversary Activities in the Information Environment	5-9
	Considerations	5-9
Appendix A	PLANNING AIDS	A-1
Appendix B	TACTICAL DECEPTION AID	B-1
Appendix C	TACTICAL OPERATIONS SECURITY AID	C-1
Appendix D	MEDIA ASSESSMENT AID	D-1
Appendix E	CONDUCTING FACE-TO-FACE MEETINGS	E-1
Appendix F	HOW TO USE TRANSLATORS	F-1
	GLOSSARY	Glossary-1
	REFERENCES	References-1
	INDEX	Index-1

Figures

Figure 1-1. Information operations capabilities 1-2
Figure 1-2. Information environment 1-3
Figure 2-1. Information operations employment 2-1
Figure 2-2. Summary of the five-step operations security process 2-3
Figure 2-3. Useful format for determining risk to critical information 2-4
Figure 2-4. Useful format for planning operations security tasks 2-5
Figure 2-5. Military deception usage 2-5
Figure 2-6. Example of a military deception work sheet 2-9
Figure 2-7. Five-step computer network operations planning process 2-17
Figure 2-8. Example message (paired to themes) 2-22
Figure 2-9. Example message development matrix 2-22
Figure 2-10. Example of a face-to-face engagement work sheet 2-25
Figure 2-11. Mutual support within information operations capabilities 2-32
Figure 2-12. Potential conflicts within information operations capabilities 2-36
Figure 2-13. Support roles of information operations, civil-military operations, public affairs, defense support to public diplomacy, and combat camera 2-39
Figure 3-1. Information operations staff estimate 3-3
Figure 3-2. Example graphic information operations estimate 3-4
Figure 3-3. Sample information operations asset list 3-5
Figure 3-4. Example of information content and flow organization 3-6
Figure 3-5. Mission analysis work sheet 3-7
Figure 3-6. Information advantage work sheet 3-9
Figure 3-7. Examples of information superiority 3-9
Figure 3-8. Effects for information operations 3-11
Figure 3-9. Tasks for information operations capabilities 3-13
Figure 3-10. Example of an information operations planning work sheet 3-14
Figure 3-11. Example information operations concept of support sketch 3-15
Figure 3-12. Format for a five-paragraph information operations annex (Army orders format) 3-16
Figure 3-13. Example of a format for a matrix information operations annex (Army orders format) 3-17
Figure 3-14. Example 1 of an information operations execution matrix 3-18
Figure 3-15. Example 2 of an information operations execution matrix 3-19
Figure 4-1. Sources of data collection 4-3
Figure 4-2. Example of an assessment graphic 4-3
Figure 4-3. Battle drill format for insurgent-related violence 4-6
Figure 4-4. Abbreviated staff battle drill 4-7
Figure 5-1. Visualization of the information environment 5-4
Figure 5-2. Example combined information overlay 5-6

Figure 5-3. Example center-of-gravity analysis and the use of the CARVER process to rank and plot critical vulnerabilities in the information environment 5-7
Figure A-1. Information operations working group duties and responsibilities A-2
Figure A-2. Sample intelligence update format A-4
Figure A-3. Sample assessment update format A-4
Figure A-4. Sample operations update format A-5
Figure A-5. Sample operations calendar A-6
Figure A-6. Information operations planning aid A-6
Figure A-7. Relationship between intelligence preparation of the operational environment and information operations A-7
Figure A-8. Sample center-of-gravity analysis A-7
Figure A-9. Sample combined information overlay A-9
Figure A-10. Mission-to-task products A-9
Figure A-11. Sample information operations mission and tasks (tactical level) A-10
Figure A-12. Sample course-of-action sketch A-10
Figure A-13. Mission analysis and information operations A-11
Figure A-14. Example of a mission-analysis work sheet A-12
Figure A-15. Example of an information operations asset/capability matrix A-13
Figure A-16. Example of a fact and assumption analysis A-14
Figure A-17. Example of a commander's critical information requirement and essential elements of information for information operations A-15
Figure A-18. Information environment A-15
Figure A-19. Information environment variances by level of war A-16
Figure A-20. Sample information environment effects matrix A-17
Figure A-21. Combined information operations overlay template A-18
Figure A-22. Example of a combined information operation A-19
Figure A-23. Relationship between center-of-gravity analysis and the planning process A-20
Figure A-24. Decisionmaking template A-20
Figure A-25. Information infrastructure template A-21
Figure A-26. Information tactics template A-21
Figure A-27. Information situation template A-22
Figure A-28. Example of an information operations estimate format A-23
Figure A-29. Graphic information operations estimate A-24
Figure A-30. Example of an execution matrix A-25
Figure B-1. Deception planning process overview B-1
Figure B-2. Deception estimate format B-2
Figure C-1. Operations security and the planning process C-1
Figure C-2. Example of a counterintelligence template C-2
Figure D-1. Media analysis process D-2
Figure D-2. Media theme assessment diagram D-3

Contents

Figure D-3. Sample media report D-4
Figure D-4. Consequence-management tracker format D-5
Figure D-5. Example of a consequence-management tracker D-5

Preface

Information operations (IO) are essential to the successful execution of military operations. The goal of IO is to gain and maintain information superiority that translates to a competitive edge in the information environment.

Users of Training Circular (TC) 18-06, *Special Forces Guide to Information Operations,* must be familiar with the decisionmaking process established in Army Doctrine Publication (ADP) 5-0, *The Operations Process,* and the operational concepts established in ADP 3-0, *Unified Land Operations.*

PURPOSE

This TC serves as a guide to describe the fundamentals of how to incorporate IO at the tactical and operational level. Appendixes A through F offer tactics, techniques, and procedures (TTP) Special Forces (SF) Soldiers can use to analyze and plan information operations. This TC implements Army and joint IO doctrine established in FM 3-13, *Inform and Influence Activities,* and Joint Publication (JP) 3-13, *Information Operations.*

This TC reinforces the definition of IO used by Army forces: IO employs the core capabilities of electronic warfare (EW), computer network operations (CNO), Military Information Support operations (MISO), military deception (MILDEC), and operations security (OPSEC), in concert with specified supporting and related capabilities, to affect or defend information and information systems and to influence decisionmaking. This TC is specifically targeted for SF; however, it is also useful to Army special operations forces (ARSOF) and the Army in understanding how SF employs IO.

SCOPE

TC 18-6 significantly affects the conduct of full-spectrum operations as an SF-common skill set that applies to offensive as well as defensive operations. This TC links to a broad variety of doctrine to provide a rudimentary understanding of IO.

APPLICABILITY

This publication applies to the Active Army, the Army National Guard/Army National Guard of the United States, and the United States Army Reserve, unless otherwise stated.

ADMINISTRATIVE INFORMATION

The proponent of this manual is the United States Army John F. Kennedy Special Warfare Center and School (USAJFKSWCS). Reviewers and users of this manual should submit comments and recommended changes on Department of the Army Form (DA Form) 2028 (Recommended Changes to Publications and Blank Forms) to Commander, United States Army John F. Kennedy Special Warfare Center and School, ATTN: AOJK-CDI-SF, 3004 Ardennes Street, Stop A, Fort Bragg, NC 28310-9610.

Chapter 1

Fundamentals

IO should be viewed as an element of combat power, focused when and where it best supports the operation. As with other elements of combat power, there is no universal formula for the application of IO. Mission, enemy, terrain and weather, troops and support available-time available, and civil considerations are the major determinants.

The purpose of IO is to achieve and maintain information superiority or advantage over the adversary at a particular time and place. To achieve an information advantage, an SF unit must understand the characteristics of the information environment in its operational area. The unit must also understand how adversary and third-party organizations use information to achieve their objectives.

Operation VALHALLA

Operation VALHALLA was a typical SF-type mission. The Jaish al-Mahdi death squad was tracked down because of the especially brutal murders of a number of civilians and Iraqi troops. On 26 March 2006, a battalion from the 10th Special Forces Group (Airborne) (SFG[A]), as part of the Combined Joint Special Operations Task Force—Arabian Peninsula (CJSOTF-AP), along with the Iraqi special forces unit it was training, engaged the Jaish al-Mahdi at their compound. The mission was successful with no friendly casualties. There were approximately 17 Jaish al-Mahdi members killed, a weapons cache found and destroyed, a badly abused hostage found and rescued, and approximately 16 Jaish al-Mahdi members detained. A combat-camera element, along with some SF Soldiers wearing helmet cameras, recorded the entire operation.

By the time the SF and Iraqi forces returned to their compound, roughly an hour after leaving the site of the firefight, someone had moved the Jaish al-Mahdi bodies. The guns of the Jaish al-Mahdi fighters were taken, and their bodies were put back inside the compound so it appeared as if the Jaish al-Mahdi members were killed while engaging in prayer. Someone then photographed the bodies in these new poses, and loaded the images onto the web, along with a press release explaining that American Soldiers had entered a mosque and killed men peacefully at prayer. This was done in under an hour. Both the American and Arab media picked up the story almost immediately. The United States (U.S.) did not release a statement until 70 hours after the operation. During the resulting investigation, which took close to a month, the SF Soldiers, who had soundly and justly defeated their adversary, were made combat ineffective by a cell-phone camera.

Paraphrased from an article written by Cori E. Dauber
Military Review
January–February 2009

Key terms used throughout this chapter are defined below:
- *IO.* The integrated employment, during military operations, of information related capabilities in concert with other lines of operation, to influence, disrupt, corrupt, or usurp the decisionmaking of adversaries and potential adversaries while protecting our own.
- *Information environment.* The aggregate of individuals, organizations, and systems that collect, process, disseminate, or act on information.
- *Information superiority.* The operational advantage derived from the ability to collect, process, and disseminate an uninterrupted flow of information while exploiting or denying an adversary's ability to do the same.

INFORMATION OPERATIONS

1-1. The possession and use of information can provide a marked advantage to one military force over another. SF units expend significant time and resources to collect, process, and internally transfer information for the purpose of mission command. Without adequate and accurate information, an SF unit is unlikely to successfully accomplish its mission or meet its objectives.

1-2. Stated in the simplest way, IO is the use of information to gain an advantage over an opponent. Such an advantage, known as information superiority, is achieved by a series of actions by military and other forces to impact both enemy forces and the operational area. To gain the advantage over the adversary, an SF unit should use any available capability at its disposal, whether doctrinal or not, to achieve information superiority at specific times and places in the operation. Figure 1-1 describes the five core capabilities, five supporting capabilities, and three related capabilities of IO. IO forces can affect data, information, and knowledge in three basic ways by—
- Taking specific psychological, electronic, or physical actions that add, modify, or remove information from the environment of various individuals or groups of decisionmakers.
- Taking actions to affect the infrastructure that collects, communicates, processes, and stores information in support of targeted decisionmakers.
- Influencing the way people receive, process, interpret, and use data, information, and knowledge.

Core Capabilities	Supporting Capabilities	Related Capabilities
Electronic Warfare (EW)	Information Assurance (IA)	Public Affairs (PA)
Computer Network Operations (CNO)	Physical Security	Civil-Military Operations (CMO)
Military Information Support Operations (MISO)	Physical Attack	Defense Support to Public Diplomacy (DSPD)
Military Deception (MILDEC)	Counterintelligence (CI)	
Operations Security (OPSEC)	Combat Camera (COMCAM)	

Figure 1-1. Information operations capabilities

1-3. Thinking about IO within the terms of this doctrinal construct may do injustice to IO's true capabilities. Field experience shows that IO is less about doctrinal capabilities than it is about understanding that every military action has the potential to positively and negatively affect populations within the operational environment. In the end, everything an SF unit or detachment does or does not do can affect the information environment, and any asset that affects information content and flow is a possible contributor to (or detractor from) achieving the mission or the commander's objective. For this reason, IO should include any methods and means that can affect information content and flow, and target perceptions and behaviors in the operating area.

THE INFORMATION ENVIRONMENT

1-4. To use IO properly, SF commanders and staffs must understand the characteristics of the information environment in their operational area. Unfortunately, visualization of the information environment is challenging because the most important aspects of the information environment—information content and flow—cannot be seen the same way we see terrain. This is because information is an abstract concept and the information environment is largely nonphysical.

1-5. The information environment has existed since humans first began communicating. That is because information resides in the minds of humans, is communicated between humans, and is the end result of how humans perceive themselves and their surroundings. To explain this phenomenon, most practitioners of IO use a three-dimensional model of the information environment (Figure 1-2).

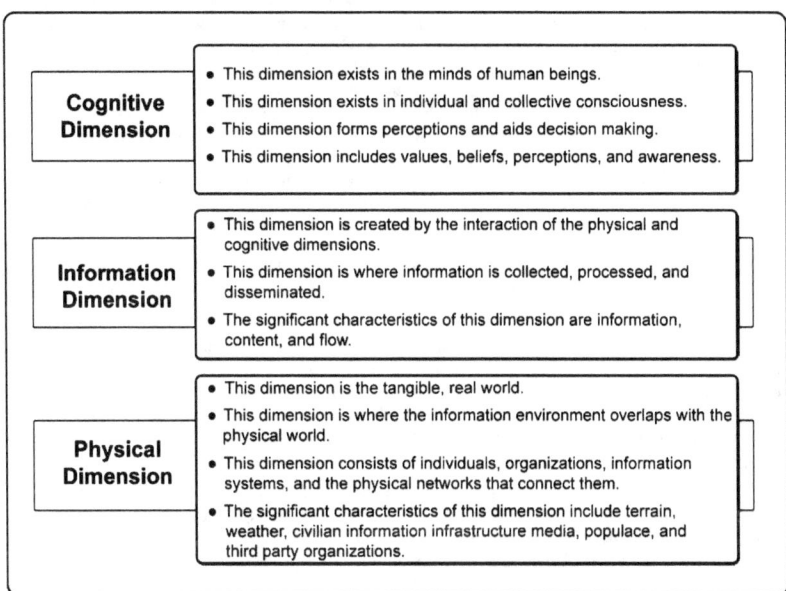

Figure 1-2. Information environment

1-6. When taken together, the information environment's three dimensions explain how the creation and flow of information causes real-world effects by converting real-world (physical) situations into human perceptions that form the basis of individual and organizational behavior. Unfortunately, although the effects of information are observable, the cause—information content and flow—is largely invisible. However, analysis of the information environment's dimensions can explain the disposition of the information environment in any specific operational area and its impact on SF operations. Broad considerations include the following:

- *The information environment is not uniform.* Physical features of the operational area (for example, terrain, information infrastructure, population demographics, and so on) determine the topography of an information environment and the cognitive aspects of the people and organizations present in the area (for example, their collective values, beliefs, and perceptions). The interactions of these factors form distinct subinformation environments, or areas in which the information environment's characteristics are notably different from those of adjacent areas.

Analysis of a specific operational area can identify subinformation environments and their effect on SF operations. SF units must anticipate having to employ IO differently within each subinformation environment.

- *Information content and flow are variable.* The relevance or importance of information changes according to the needs of the various population groups and organizations. For example, people located in an area devastated by a natural disaster desire information concerning humanitarian assistance, whereas the populace in an insurgent-infested area is primarily interested in information related to security. The task for SF units is to determine what information in the operational area is important to the mission and then to identify and track its primary themes and flow, just as SF units observe and monitor the presence of enemy forces.

- *The information environment's character changes by the level of war and mission.* The information environment becomes less tangible and more conceptual as operations move from the tactical to the strategic. At the tactical level, information flow is primarily by short-range communications systems and observable means, such as graffiti and banners. What people see of their physical surroundings is critical to their situational awareness, perceptions, and behavior. On the other end of the spectrum—the strategic level—the information environment is impacted less by physical features and more by abstract ideas, ideologies, and philosophies. Information flow is not terrain-dependent, extending well over the horizon by long-range and mass-communications systems. Finally, the assigned mission (for example, combat, peacekeeping, humanitarian assistance, and so on) is a critical determinant of an SF unit's relationship to its information environment because it establishes the relative importance of the information environment's specific characteristics to the conduct of operations. For example, in conventional combat, the physical information infrastructure in the operational area is often a dominant characteristic because of its potential use by the enemy. In counterinsurgency missions, populace support (a cognitive aspect of the information environment) is a critical characteristic because of its importance to enemy and friendly operations.

1-7. To impact the information environment, an SF unit must identify subinformation environments and information nodes in its operating area. Subinformation environments are areas in which the information environment's characteristics and effects are notably different from those of adjacent areas. Information nodes are places, persons, or infrastructure that shape information content and flow by creating or transmitting information into the surrounding area. It is important to note that information nodes can change from day to day so what worked one day may not necessarily work the next.

1-8. Operations in the information environment are asymmetrical and not benign, often favoring one side over another. Opposing forces use the information environment just as they use the physical environments of air, land, and sea to place their enemy at a disadvantage and to achieve their objectives. Furthermore, U.S. adversaries do not use the information environment in the same way or have the same means as U.S. forces. Understanding this, an SF unit must identify how its adversary views and uses the information environment. This is a challenge, because even though two opposing forces occupy the same operational environment, they will not have the same capabilities in the information environment. It is important to avoid *mirror imaging* U.S. concept of IO upon the adversary and mismatching U.S. capabilities and vulnerabilities to those of the adversary. Chapter 5 provides additional information.

INFORMATION SUPERIORITY

1-9. Information superiority is the purpose of IO. It is also the reason why a commander allocates resources to IO. Information superiority should not be treated as a doctrinal catch-phrase. Just as each mission's end state is different, so is information superiority. For example, during combat operations, information superiority can be gaining surprise over the enemy or preventing the enemy from employing its reserve forces. During counterinsurgency operations, information superiority can be gaining populace support for friendly operations or preventing enemy freedom of flow. In each case, information superiority is defined specifically for the mission in terms of what advantage is sought for the friendly force.

1-10. To achieve information superiority, an SF unit uses information to actively attack the adversary and to shape the information environment to the force's own advantage. This duality of operations—attacking the adversary and shaping the information environment—is analogous to "fires and maneuvers," where fires equate to attacking the adversary's ability to use information, and maneuvers are actions to seize and retain information nodes to gain a positional advantage in the information environment. To be effective, an information operation balances lethal and nonlethal activities to attack the adversary with those that shape the information environment. Through a combination of both, an SF unit seeks information superiority over its opponent.

1-11. An SF unit will rarely achieve absolute and universal information superiority. The actions of opposing forces, as well as the information content and flow in the operational area, are not static. Therefore, information superiority is a localized and transitory condition over the adversary. SF units seek information superiority at certain times and places, usually at or before the decisive point of the operation. Chapter 3 provides additional information.

INFORMATION OPERATIONS CAPABILITIES

1-12. SF operations are not planned for the purpose of using any particular capability. Mission requirements, namely campaign objectives, operating environments, and adversary and friendly forces, dictate what capabilities a commander uses and how they are employed. IO are no different.

1-13. Although often described as a discrete set of capabilities (doctrinally organized as core, supporting, and related capabilities), IO are much more than that. Capabilities used for information operations should be selected based on mission requirements, specifically focused on the desired effect. Some doctrinal IO capabilities—MISO, EW, and CNO—require trained specialists and equipment. However, each element of an SF unit must be able to employ OPSEC, MILDEC, and COMCAM, as well as other IO enablers, such as key-leader engagements and rewards programs. Additionally, various IO capabilities are used in concert with PA, Civil Affairs (CA), host nation (HN), foreign internal defense (FID), partner nation information capabilities, and select interagency capabilities (for example, provincial reconstruction teams). Chapter 2 provides additional information leverage.

CONSIDERATIONS

1-14. Subject-matter experts for core IO capabilities are typically positioned at the special operations task force (SOTF) level and higher when deployed; however, it is essential that commanders at detachment levels understand the core capabilities of IO and how to effectively utilize them to achieve information superiority and accomplish the assigned missions. SOTF staffs can include an IO planner, EW planner, Military Information Support (MIS) planner, a MIS detachment commander, a CA planner, and a COMCAM and PA representative. At the joint special operations task force (JSOTF) level, the staff mirrors the battalion with the addition of a special technical operations planner and the group public affairs officer (PAO), along with retaining the conduits to leverage the higher headquarters (HQ) assets and interagency capabilities. The subject-matter experts at the SOTF and JSOTF can provide training and recommendations to the Special Forces operational detachments A (SFODAs) and Special Forces operational detachments B (SFODBs) on how best to utilize information capabilities, and provide assistance in coordination and deconfliction for IO capabilities in support of their concept of operations.

CONCLUSION

1-15. IO is the use of information as a military capability. Many of the principles and concepts that guide the conduct of other military operations also guide the employment of IO. One way for SF commanders and staffs to integrate IO into operations is to consider IO in terms of the factors of mission, enemy, terrain and weather, troops and support available-time available, and civil considerations:
- *Mission.* The role of IO in the unit mission is to achieve information superiority. As such, it is important to identify exactly what advantage over the enemy IO is expected to achieve.
- *Enemy.* Gaining an advantage over the adversary in the information environment starts with pairing friendly capabilities and vulnerabilities in the information environment against those of the adversary. An information operation that defeats the adversary's capabilities and turns the information environment to the friendly force's favor will achieve information superiority.

- *Terrain and weather.* Terrain and weather interact with the information environment to affect information content and flow as well as the employment of IO capabilities. SF units must adjust the employment of IO to the terrain and weather.
- *Troops and support available.* Rarely does an SF unit have all the assets needed to conduct an information operation. Commanders and staffs can fill the gap by thinking beyond doctrine for other ways and means to affect information content and flow.
- *Time available.* Regardless of echelon, IO requires long lead times compared to other operations. Typically, IO must be planned one phase or event in advance of fire and maneuver.
- *Civil considerations.* When civil considerations are important to the unit mission, IO capabilities can be applied to influence the populace, if doing so will achieve an advantage over the adversary.

Chapter 2

Information Operations Capabilities and Tactics

This chapter focuses on the employment of IO capabilities and tactics to gain information superiority. It also shows the links among the capabilities in diagram form. The core and supporting IO capabilities are similar to the warfighting functions. They are independent capabilities that, when taken together and synchronized, constitute IO. IO planners must not let doctrine constrain their selection of capabilities. Any available assets, means, capabilities, or tactics that can shape the information environment or target the adversary's ability to use information should be considered for employment as part of IO.

The use of assets and means for the purposes of IO require judgment in application. Some capabilities—notably MISO, EW, and CNO—are disciplines that require specialized training and skill sets. Employment of these capabilities requires specialized technical expertise to properly plan and execute. Other capabilities already reside within a command or unit and require only planning and coordination to employ them as part of IO. Figure 2-1 outlines some of the more commonly employed IO capabilities.

	Capability	Employment
Doctrinal Capabilities	OPSEC	Deny critical friendly information to the adversary.
	MILDEC	Mislead adversary leaders into making decisions that are favorable to friendly forces.
	MISO	Change or reinforce attitudes and behavior favorable to friendly objectives.
	EW	Degrade, disrupt, or deny adversary use of the electromagnetic spectrum (EMS).
	CNO	Degrade, disrupt, or deny adversary use of cyberspace.
	COMCAM	Visually document friendly and adversary forces' operations and activities.
	CMO	Gain local populace acceptance and support.
	PA	Inform populace groups and counter misinformation and propaganda.
Tactics	COMCAM	Visually document friendly and adversary forces' operations and activities.
	Local populace and key-leader engagements	Gain support for friendly-force operations and HN government activities.
	Countering adversary information	Neutralize hostile propaganda or mitigate its effects.
	Rewards program	Influence adversary leaders' perceptions.

Figure 2-1. Information operations employment

OPERATIONS SECURITY

2-1. OPSEC is a universal IO capability. It is not just an "in-garrison" competency and needs to be operational at strategic, operational, and tactical levels. OPSEC should be included in all plans, operations, and activities. The goal of OPSEC, in conjunction with unit security programs, is to achieve essential secrecy. Essential secrecy is concerned with the content and flow of critical information. Military forces seek critical information about their opponents to fulfill their own information needs. To do this they attempt to collect accurate, timely, and relevant information, process the information, and disseminate it for use in planning and directing operations. Conversely, if a military force is to prevent its adversary from gaining useful information, then it must prevent the flow of critical information from friendly to adversary forces. At its core, OPSEC is an approach to conducting operations. To have a good OPSEC program, it is imperative that the unit identify this critical information, understand the OPSEC indicators and vulnerabilities, and practice essential secrecy. JP 1-02, *Department of Defense Dictionary of Military and Associated Terms*, defines essential secrecy as *the condition achieved by the denial of critical information to adversaries*. Essential secrecy depends on the combination of two approaches to protection—security programs to protect classified information, and OPSEC to deny adversaries critical information (which is often unclassified).

2-2. Each command and operation has a tremendous amount of information, both classified and unclassified, that must be protected. However, denying all information about a friendly operation or activity is seldom cost-effective or realistic. Central to this idea is the concept of essential secrecy. By achieving essential secrecy, military forces protect their intentions, capabilities, and activities to retain initiative and the element of surprise for operations. As a condition, essential secrecy is not static—it must first be developed and then maintained as the situation and mission evolve. Essential secrecy cannot be achieved in all places and at all times; therefore, the protection of information must be focused and prioritized to counter specific threats.

2-3. Essential secrecy and the protection of critical information is not the exclusive responsibility of OPSEC. It is the result of mutually supportive OPSEC and security programs. The purpose of OPSEC is to prevent, or at least limit, the flow of sensitive, unclassified information to adversary forces. The actual content of the information, whether classified or unclassified, is the responsibility of information security program controls and procedures. OPSEC denies critical, friendly information to the adversary by eliminating or reducing to an acceptable level the vulnerabilities of friendly actions to adversary exploitation. Because OPSEC is not the sole contributor to essential secrecy, an IO objective can integrate other capabilities—such as MILDEC, physical security, information security, and CI—that are not related to OPSEC.

2-4. OPSEC is a process of identifying and protecting critical information and actions that could benefit the adversary. A good OPSEC operation starves the adversary's intelligence system by denying it the information it seeks. Without information on friendly organization, disposition, and intent, the adversary leader's decisionmaking is degraded.

2-5. The basis for OPSEC's contribution to an operation is the commander's key tasks for IO. This means that, for OPSEC to be part of an information operation, at least one essential IO task should address the protection or defense of friendly information.

2-6. Although the purpose of OPSEC is a constant, its focus may change by echelon. At the tactical level, OPSEC prevents the adversary's detection and identification of friendly activities and operations to prevent the targeting of critical assets and countering of current activities and operations. Operational-level OPSEC prevents the disclosure of intentions, capabilities, and future operations (that is, courses of action [COAs]) to avoid the compromise of planning and operations. Tactical-level OPSEC addresses specific measures to defeat the adversary's collection capabilities; whereas, at the operational level, OPSEC addresses broad guidance or general measures for the entire force and new measures to counter the adversary's intelligence capabilities.

2-7. As a way to systematically identify, analyze, and protect critical information relevant to the mission, OPSEC is integrated into the military decisionmaking process. Figure 2-2, pages 2-3 and 2-4, depicts a summary of the five-step OPSEC process.

Note. The five-step process is used at the JSOTF where the assets to form an OPSEC working group exist. At the tactical level, the type of information contained in the OPSEC work sheet (Figure 2-3, page 2-4) and the OPSEC tasks (Figure 2-4, page 2-5) needs to be considered when developing a concept of operations.

1. **Identify Critical Information.** Determine what information needs protection by identifying the information required by the adversary to prevent friendly-force mission success (list of critical information or essential elements of friendly information [EEFI]). Adversaries can derive critical information from the aggregation of indicators resulting from the observation or detection of friendly-force activity. Friendly actions generate indicators (detectable actions and open-source information) that can be collected and developed into critical information (facts about friendly intentions, capabilities, and activities). An adversary can plan and execute its own operations by using critical information. To identify critical information—
    - Identify what information is critical to the friendly mission. Sources of critical information include higher HQ plans and operations orders, commander's guidance, and current unit-critical information lists. Focus on friendly-force intentions (time and place of units and operations), capabilities, and vulnerabilities (strength, technologies, and tactics).
    - Keep in mind that critical information is different for every operation. Do not use a "cookie-cutter" approach. Continually develop or refine critical-information lists.
    - Use an OPSEC working group to take advantage of subject-matter experts (for example, aviation, communications, and computer systems).
    - Identify the length of time each element of critical information must be protected (not all information needs protection for the duration of the operation).
    - Write critical information in the form of a statement (do not write critical information in the form of a question). Generic examples include current and future locations of unit elements; intelligence, surveillance, and reconnaissance capabilities and limitations; and unit movement methods and routes.
    - List the elements of critical information (for example, time and route of helicopter flight) in the order of priority and keep to a manageable number (perhaps five).
2. **Analyze Threat.** Identify the threat to the critical information by determining the adversary's information needs and collection capabilities:
    - Information needs are items of information the adversary requires. Do not bother trying to protect information that the adversary already has.
    - Collection capabilities include human intelligence (HUMINT), signals intelligence (SIGINT), imagery intelligence, and open-source intelligence (OSINT). An estimated 90 percent of the adversary's information needs are met from OSINT.

    Example threat analysis: Adversary knows: personnel and equipment move by helicopter; adversary needs: departure times and routes of flight; adversary collection method: visual observation by spotters.
3. **Analyze Vulnerabilities.** Identify each element of critical information and its vulnerability to adversary intelligence collection. These are known as OPSEC vulnerabilities and are the result of detectable indicators of the critical information. OPSEC indicators become OPSEC vulnerabilities if they can be observed, analyzed, and acted upon by the adversary. To determine OPSEC vulnerabilities:
    - Identify OPSEC indicators. Determine what detectable actions and OSINT can be interpreted or pieced together by the adversary to derive the unit's critical information.
    - Compare OPSEC indicators to adversary collection capabilities. Determine which indicators can be observed, analyzed, and acted upon by the adversary.

    Example OPSEC vulnerabilities: Direction of flight, helicopters taking off, loading of troops and equipment, and assembly of troops and equipment.
4. **Assess Risks.** The goal is to reduce risk to an acceptable level based on the commander's guidance. Conduct a risk assessment for each vulnerability to determine which really need protection. Focus on the

Figure 2-2. Summary of the five-step operations security process

Chapter 2

vulnerabilities that produce the most risk to mission success and, therefore, are an unacceptable risk, and then select one or more OPSEC measure for each vulnerability:
- There are three types of OPSEC measures:
  - Action controls that change unit procedures, activities, and actions (randomized routine activities, avoid repetitive tactics and procedures).
  - Countermeasures that disrupt enemy information gathering and targeting (jamming [EW], physical attack, and camouflage, cover, and concealment).
  - Counteranalysis that deceives the enemy by providing false indicators (decoys, deception in support of OPSEC).
- Decide which OPSEC measures to implement. Check that OPSEC measures do not create new vulnerabilities. Balance OPSEC measures with operational effectiveness (risk versus unit resources). Developing OPSEC measures is a balance between cost and resources in terms of time, personnel, assets, and interference with operations.

Example OPSEC vulnerabilities and mitigating measures: direction of flight (fly in false direction, change direction en route), helicopters taking off (vary flight times, conduct false missions), loading of troops and equipment (load just prior to takeoff), assembly of troops and equipment (assemble troops and equipment under cover).

5. **Apply OPSEC Measures.** Tasks turn OPSEC measures into specified actions. Because OPSEC measures do not follow any doctrinal format, it is necessary to convert them to tasks that the executing units and elements can understand. Develop tasks that support the command's key IO tasks, as well as protect and control the specific indicators associated with key operational tasks:
- Rewrite approved OPSEC measures as tasks. A useful format is task, purpose, and method. In general, for OPSEC, a *task* is an action that controls or protects observable activities, *purpose* can be critical information requiring protection, and *method* is the OPSEC means or methods used to execute the task.

Example OPSEC task that supports combat operations: task—jam enemy ground surveillance radars, purpose—conceal flow of combat elements from electronic collection, method—screen jamming.

Example OPSEC task to support stability operations: task—deny civilian populace access to base-camp overwatch sites, purpose—prevent line-of-sight observation of security activities, method—unit patrols, local police.

- Assign responsibility and coordinate OPSEC tasks with units and staff, to include the intelligence directorate of a joint staff (J-2)/assistant chief of staff, intelligence staff section (G-2)/intelligence staff officer (S-2), and CI for monitoring, and then include OPSEC tasks in the operation plan or operation order.

Figure 2-2. Summary of the five-step operations security process (continued)

EEFI	Vulnerability	Indicators	Adversary Collection	Risk Level	OPSEC Measure	Residual Risk	Assess
Location of unit elements	Assault-force insertion	Rotary-wing movement	Sympathetic populace	Extremely High	False insertions	Medium	No adversary contact on ground flow
		Ground movement	Direct observation	High	Reconnaissance element placed on route	Low	Adversary surprised on objective

Figure 2-3. Useful format for determining risk to critical information

EEFI	Adversary Collection	Vulnerable Indicators	Standing Operating Procedures or Current OPSEC Measures	Additional OPSEC Measures	OPSEC Tasks
Task organization	Spotters on forward operating bases and main supply routes	Vehicle markings	Cover vehicles markings	Remove unit markings	1st Battalion 2d Battalion
		Command vehicles	None	No unsecured communications	Group HQ

Figure 2-4. Useful format for planning operations security tasks

2-8. During operations, the current operation staff should monitor and adjust the elements of critical information based on the adversary's reaction to the implemented OPSEC tasks and for inadvertent disclosure by friendly forces. Tools useful to planning and implementing an OPSEC plan are the OPSEC working group, OPSEC standing operating procedures (SOPs), and OPSEC work sheets.

2-9. The OPSEC working group is a group of subject-matter experts that determines critical information, identifies OPSEC vulnerabilities, coordinates and synchronizes OPSEC measures and tasks, and assesses the effectiveness of OPSEC tasks. Typical membership includes an intelligence analyst to assist with threat analysis, CI personnel to analyze vulnerabilities, a force protection officer, communications and aviation representatives, and subordinate unit liaison officers. The OPSEC working group should conduct periodic assessments of command critical information, threat collection capabilities, OPSEC vulnerabilities, and OPSEC measures.

2-10. An OPSEC SOP is critical to ingraining OPSEC into unit operations. The SOP should be short and direct and should include standing critical information or EEFI, standing OPSEC measures, composition and responsibilities of the OPSEC working group, and OPSEC assessment procedures.

MILITARY DECEPTION

2-11. JP 1-02 defines MILDEC as *actions executed to deliberately mislead adversary military decision makers as to friendly military capabilities, intentions, and operations, thereby causing the adversary to take specific actions (or inactions) that will contribute to the accomplishment of the friendly mission.*

2-12. MILDEC is more of a process or way of thinking than a capability with tangible assets and resources. It may be executed using a unit's own troops and equipment. An effective deception does not have to be elaborate or complex; however, any time deception is part of an operation, it is the main effort for the information operation and should be included in the defined operational advantage (information superiority) provided for the mission.

2-13. MILDEC is a method, not a result. MILDEC is not conducted merely to deceive an adversary. Deception is used only to support the mission. Figure 2-5 shows ways to employ MILDEC.

Application	Purpose	Focus
MILDEC	Achieve an exploitable advantage	The adversary's leaders and decisionmakers
Deception in support of OPSEC	Deny information about friendly forces	The adversary's intelligence, surveillance, and reconnaissance capabilities
Deception as part of camouflage, concealment, and decoys	Protect units, systems, and personnel	The adversary's weapons and target-acquisition system

Figure 2-5. Military deception usage

Chapter 2

2-14. MILDEC actively targets adversary leaders and decisionmakers in support of specific battles and engagements. It creates an exploitable advantage by misleading or confusing the adversary's decisionmaker. Distorting, concealing, or falsifying indicators of friendly intentions, capabilities, or dispositions that the adversary will see and collect can mislead or confuse the adversary. MILDEC is conducted at all levels—strategic, operational, and tactical—and must be carefully coordinated to deconflict operations between the HQ and subordinate units.

2-15. Deception in support of OPSEC is conducted to reinforce unit OPSEC and is planned using the OPSEC plan as the basis for the deception. A deception in support of OPSEC uses false information about friendly forces' intentions, capabilities, or vulnerabilities to shape the adversary's perceptions. It targets the adversary's intelligence, surveillance, and reconnaissance abilities to distract the adversary's intelligence collection away from, or provide cover for, unit operations. A deception in support of OPSEC is a relatively easy form of deception to use and is very appropriate for use at battalion-level and below. To be successful, a balance must be achieved between OPSEC and MILDEC requirements.

2-16. Camouflage, concealment, and decoys are normally individual or unit responsibilities and governed by SOP. These actions may be taken for their own ends. They can also play a role in a larger MILDEC or deception in support of OPSEC operations where camouflage, concealment, and decoys comprise just a few of many elements that mislead the adversary's intelligence, surveillance, and reconnaissance abilities. Merely hiding forces may not be adequate, as the adversary may need to "see" these forces elsewhere. In such cases, cover and concealment can hide the presence of friendly forces, but decoy placement should be coordinated as part of the deception in support of OPSEC.

2-17. The uncertainties of combat make decisionmakers susceptible to deception. The basic mechanism for any deception is either to increase or decrease the level of uncertainty (commonly referred to as ambiguity) in the mind of the deception target. Both MILDEC and deception in support of OPSEC present false information to the adversary's decisionmaker to manipulate their uncertainty. Deception may be used in the following ways:

- *Ambiguity-decreasing deception.* This type of deception presents false information that shapes the adversary decisionmaker's thinking so he makes and executes a specific decision that can be exploited by friendly forces. This deception reduces uncertainty and normally confirms the adversary decisionmaker's preconceived beliefs so the decisionmaker becomes very certain about his COA. By making the wrong decision, which is the deception objective, the adversary could misemploy forces and provide friendly forces an operational advantage. For example, ambiguity-decreasing deceptions can present supporting elements of information concerning a specific adversary's COA. These deceptions are complex to plan and execute, but the potential rewards are often worth the increased effort and resources.

- *Ambiguity-increasing deception.* This deception presents false information aimed to confuse the adversary decisionmaker, thereby increasing the decisionmaker's uncertainty. This confusion can produce different results. Ambiguity-increasing deceptions can challenge the enemy's preconceived beliefs, draw enemy attention from one set of activities to another, create the illusion of strength where weakness exists, create the illusion of weakness where strength exists, and accustom the adversary to particular patterns of activity that are exploitable at a later time. For example, it can cause the target to delay a decision until it is too late to prevent friendly-mission success. It can place the target in a dilemma for which there is no acceptable solution. It may even prevent the target from taking any action at all. Deceptions in support of OPSEC are typically executed as this type of deception.

2-18. Before planning a deception, it is first necessary to determine if there is a deception opportunity. A deception may be a feasible option if it is appropriate to the mission and if there is a possibility of success against the adversary. The following questions should be considered when planning deception:

- *Is the adversary susceptible to deception?* Planners should use the J-2/G-2/S-2 adversary COA as a basis to develop information about the adversary's system and decisionmaking process. Planners should determine how the deception target acquires and acts on information, what knowledge the target has of the situation and how the target views the friendly force. If

necessary, planners should make assumptions. To do this, they should try to place themselves in the position of the adversary without mirror imaging.
- *Does the friendly mission lend itself to deception?* Some missions are better suited to deception than others. Planners should not feel compelled to work deception into every operation. Generally, when a unit has the initiative and can exercise some control over the mission area of operations (AO), then deception is possible.
- *Do constraints prevent the use of deception?* Other than the constraints imposed by authorities and political considerations, the most important consideration is time. Execution of the mission must allow enough time for the adversary to see the deceptive activities, reconstruct the activities into the deception story, form the desired perceptions, and issue the orders that will cause the adversary force to act in a manner consistent with the deception objective.
- *Are friendly assets available?* To successfully deceive the adversary, MILDEC requires assets. However, very few assets are specifically designed and designated for deception purposes. This means that existing assets have to dedicate support to the deception. This is sometimes difficult, especially when assets are limited. Therefore, the unit may have to be creative to find assets and to use them efficiently.

Note. JSOTF will support strategic MILDEC plans and plan operational MILDEC. At the SOTF and below, units will use tactical MILDEC. Appendix B provides additional information and a tactical deception aid format.

2-19. As with other operations, deception planning follows the military decisionmaking process. Planning a deception does not have to be difficult, but there are certain steps that must be taken to ensure the deception is properly constructed. The steps are as follows:
- *Determine the deception goal.* The deception goal is the desired contribution of the deception to friendly-mission success. In other words, what advantage does the deception provide for friendly forces (for example, provide target opportunities for friendly forces)?
- *Determine the deception objective.* The deception objective is the purpose of the deception operation expressed in terms of what the adversary is to do or not to do at the critical time and location. In simpler terms, it is the action or inaction that friendly forces want the adversary to take (for example, cause insurgent forces to move into the open).
- *Identify the deception target.* The target is the adversary decisionmaker with the authority to make the decision that will achieve the deception objective (for example, the insurgent group commander).
- *Identify desired perceptions.* These are what the deception target must believe to make the decision that will achieve the deception objective. Based on the deception objective and target, the planner must determine the nature of the desired perceptions—will they increase or decrease the target's uncertainty (ambiguity increasing or decreasing)? Desired perceptions eventually translate to resource requirements; therefore, the number of perceptions should be kept to an absolute minimum to conserve the assets needed for the deception (for example, U.S. forces are going to attack from the south).
- *Develop the deception story.* The deception story is a plausible, but false, view of the situation which leads the deception target to act in a manner that accomplishes the deception objective. To be plausible, the story must be integrated into the overall COA. The story is built and stated exactly as the planner wants the target to reconstruct it. To develop the deception story, the planner thinks about how the target sees the situation and then writes the story like the deception target's own intelligence estimate. The story is always written from the target's perspective—what does the target expect to see and think and what will he do (for example, indications are that U.S. forces are massing to the south in preparation for an attack)?
- *Identify the deception means.* These are the methods, resources, and techniques that the unit will use to create required observables (things the adversary decisionmaker needs to see to deduce the desired perceptions) and act out the deception story. The planner must determine for each desired perception what means—physical, technical, and administrative—can be used. Physical means

are observable physical activities of forces, systems, and individuals that present visual indicators. Technical means include radio broadcasts, radar emissions, and electromagnetic deception. Administrative means are used to convey oral, pictorial, documentary, or other material evidence to the deception target.

- *Develop deception events.* These are the activities conducted by the deception means at a specific time and location to convey the deception story to the target. To convey the deception story, the deceptive activities must be observed by the adversary. To determine this, planners pair up the available deception means with the capabilities of the adversary's intelligence collection system. If the adversary intelligence system can "see" the deceptive event, then it can collect the information it needs to piece together the deception story. Deception events must be translated into tasks to subordinate units if the deception operation is to be executed (for example, loudspeaker simulating vehicle traffic, SFODA present in area).
- *Develop OPSEC measures.* Without OPSEC to deny critical information to the adversary, the deceptive activities may not convince the adversary to believe the deception story. In order for the deception to be successful, the unit must adhere to a strict need-to-know policy.
- *Develop assessment requirements.* Collecting feedback is a difficult challenge. However, to judge the effectiveness of the deception, it is necessary to have indications of how the target is responding to the deception. Ideally, there will be indicators of whether the target is receiving the deception story as planned, and if the target is acting in accordance with the deception objective (for example, insurgents move from building to highway).
- *Develop a termination plan.* A deception operation does not just end on its own. Part of the operation is a termination plan that establishes when organized deception activities cease, and how deception means, techniques, and events will be protected. This is important, because there is no logic in executing a deception after the objectives have been met. Additionally, the adversary should not know what deception means, techniques, and events were used. Otherwise, the next deception operation may not have the desired effect due to the adversary gaining insights into friendly TTP.

2-20. In time-constrained deception operations, the "see–think–do" methodology can be used as an abbreviated planning process. The planner uses this process by identifying what he wants the target to *do* (for example, the deception objective), then determines what the target must *think* (for example, required perceptions), and then establishes what the target must *see* (for example, deception events).

2-21. Deception operations cannot proceed without approval or coordination. Two authorities can direct a deception operation: a higher HQ and the unit commander. In both cases, the command's deception plan must be coordinated with the higher HQ. To ensure coordination, deception plans are normally approved two levels above the employing unit HQ. It is imperative that a deception be thoroughly coordinated to prevent information fratricide; that is, employing deception in a way that causes effects in the information environment that impede the conduct of friendly operations or adversely affect friendly forces.

2-22. The military deception work sheet (Figure 2-6, pages 2-9 through 2-11) is a tool that can be used to capture the key elements of the deception plan. Elements of planning information are listed on the work sheet in the order they are developed using the deception methodology.

Information Operations Capabilities and Tactics

Unit Mission: *(Concise restatement of the mission. Identifies the operational goal(s) of the command to which the deception must contribute.)* Establish a bridgehead across the Knewt Canal to rapidly advance forces deep into Towie territory and occupy key terrain on Sangria Ridge.	**Deception Constraints:** *(Identify constraints on the command from higher to lower as they affect the deception plan.)* All liaison activities with members of friendly foreign governments and Allies that concern this operation will not be conducted with prior approval by the Atari Chief of Staff of the military. War stocks, particularly ammunition, will be stored, handled, transported, and issued with exercise ammunition. The perception that Atari is conducting Exercise SACRED HEART must be maintained. Wartime SOPs, wartime communications, and wartime modes of operation will not be implemented until two hours prior to commencement of the actual attack operation.

Commander's Guidance for Disposition/Deception Goal: *(Describe the desired effects or the end state a commander wishes to achieve [commander's intent for the deception operation].)*

Intent: We will mask the plans for our attack and build-up of our military strike forces by massing supporting logistic elements, the forward deployment of our assault engineers and air defense umbrella, and our increased communications under the guise of Exercise SACRED HEART (the annual Atari command post and maneuver exercise). Exercise SACRED HEART provides the overall cover to conceal our actual intentions.

End State: Numerically and qualitatively superior combat, combat support, and combat service support elements of the Atari 1st Army will be fully deployed on the south bank of Knewt Canal to execute canal crossing operations, drive deeply into Towie territory, and occupy key terrain on Sangria Ridge.

Atari Commander's Goal Statement: Use MILDEC to achieve operational surprise during Atari canal crossing operations; and enable Atari freedom of maneuver during our drive to the Sangria Ridge.

Deception Objective: *(Describe the desired action or inaction on the part of the adversary at the critical time and location.)*

Cause the Towie front commander to delay mobilization and commitment of the Towie strategic reserve in response to operations. (Note: If a delay in mobilization and commitment of the strategic reserve is achieved, this delay then contributes to/supports the Atari commander's goal statement of achieving surprise.)

Deception Target: *(Identify adversary decisionmakers responsible for the actions(s) or inaction(s) specified in the deception objectives.)*

The Towie front commander has sole authority to mobilize and commit the Towie strategic reserve—he is the single and only decisionmaker that can make this decision.

Desired Perceptions: *(Describe what the deception target must believe for it to make the decision that will achieve the deception objective.)*

The Towie front commander must believe:

  - The Atari military is not preparing for immediate combat operations against the Towie forces or nation.

  - The Atari military is conducting Exercise SACRED HEART to improve wartime fighting efficiency, mission command, and logistics; hence, the buildup of its military forces.

  - Towie intelligence should detect visible signs of impending combat operations, but Atari forces would be preparing for an exercise and training in the open.

Deception Story: *(Outline a scenario of friendly actions or capabilities that will be portrayed to cause the deception target to adopt the desired perception.)* Atari military forces are conducting Exercise SACRED HEART. The purpose of Exercise SACRED HEART is to improve the fighting efficiency, mission command, and logistics of the Atari Army via the exercise's numerous training and maneuver phases.	**Deception Means:** *(Describe how the plan will be implemented and how it supports the unit's overall mission.)* Atari forces can employ the following MILDEC means to get the target to take the action desired: physical, technical, and administrative. These means can be employed independently or in collaboration depending on the situation. *(continued)*

Figure 2-6. Example of a military deception work sheet

Chapter 2

	Physical means are those activities and resources used to convey or deny selected information to the decisionmaker. Physical means include operational activities and resources such as: - The movement of the Atari Army and Air Force. - Exercise SACRED HEART subexercises and training events. - Atari logistic actions and the location of stockpiles and repair facilities during Exercise SACRED HEART. - Reconnaissance and surveillance activities performed during Exercise SACRED HEART. ***Technical means*** are the military material resources and their associated operating techniques used to convey or deny selected information to an adversary. As with any use of Atari military material resources, any use of technical means to achieve MILDEC will strictly comply with Atari domestic and international law. A variety of technical means include the following: - Deliberate radiation energy is accomplished when Atari command posts, reconnaissance and surveillance, and air defense networks go operational during Exercise SACRED HEART. - Atari multimedia (radio, television, sound broadcasting, or computers). ***Administrative means*** include resources, methods, and techniques designed to convey or deny oral, pictorial, documentary evidence. The best example of this is a well-publicized announcement (all media outlets) of the upcoming Exercise SACRED HEART.
Assessment: *(Describe the methodology to assess the deception plan if the plan is successful; if the deception fails, or is compromised to Allies or adversaries.)* The command military deception officer will use feedback and intelligence information collected by the command's intelligence officer/directorate to assess if the deception plan is successful, has failed, or had been compromised. Feedback: Is the information providing indications of the response (positive or negative) of the deception target and conduits to elements of the deception. Target (Analytical) Feedback: This is information or analytical determinations regarding the actions of the target in response to the deception executed by the deceiver. Conduit (Operational) Feedback: This is information that provides indications of if and how the conduits are receiving, processing, and transmitting elements of the deception to the target. Indications of identification of enemy deception and counterdeception are provided by the command's intelligence officer/directorate. It is their responsibility to identify foreign deception operations against friendly forces.	**OPSEC Measures to Protect Deception:** *(What OPSEC and other countermeasures are to be used to protect the deception plan from compromise.)* OPSEC measures to implement include the following: - The true nature of the operation will be strictly enforced by using the principles of "need to know." - Encrypted communications, secure land lines, and couriers will be used to convey information about our true intentions. All other traffic will use the Exercise SACRED HEART communications network to pass information. - The Exercise SACRED HEART SOP is in effect command wide; implement the wartime SOP during Exercise SACRED HEART pause but no later than two hours prior to commencement of the actual operation. - No special emphasis will be placed on camouflage and concealment of actual assault forces. In the vicinity of crossing sites standard camouflage and concealment practices to be in place and enforced. - Standard counterintelligence operations are executed throughout the exercise area but with increased focus on the crossing site areas. - OPSEC assessments and monitoring will be in effect up until the time of the attack.

Figure 2-6. Example of a military deception work sheet (continued)

Information Operations Capabilities and Tactics

	Termination Plan: *(What is the plan for terminating he deception plan if the desired effects are achieved, not achieved, or compromised?)*
	Reasons for Termination:
	- Success. *The deception operation has run its course and the MILDEC operation concludes because the deception target (Towie front commander) took the action envisioned in the MILDEC plan.*
	- Change of mission scenario. *The overall operation situation has changed and events and circumstances that prompted the deception operation no longer pertain or are applicable and the MILDEC operation is terminated by executing commander.*
	- Recalculated Risk/Success/Probability of Success Scenario. *Key elements in the deception have changed in a negative way that increases the risks and costs (for example, casualty estimates rise) to the commander and the commander elects to end the deception.*
	- Failure Scenario. *The deception target does not understand key elements of the deception or does not care about the deception elements being executed and therefore he fails to take the action envisioned in the MILDEC plan. The MILDEC plan is terminated because the Towie front commander has not taken the bait.*
	- Compromise. *The deceiver believes the adversary has learned some or all elements of t he actual deception operation. Compromise of the deception poses special challenges to the deceiver and the termination process. Not only can important deception capabilities and techniques be placed at risk, but once the adversary has discovered the MILDEC operation he may be able to "read the evidence" and reconstruct the tails of the MILDEC. Worst case scenario—the enemy (Towie front commander) may be able to exploit the compromised deception by initiating his own counterdeception operation to counter/negate the true operation.*

Figure 2-6. Example of a military deception work sheet (continued)

MILITARY INFORMATION SUPPORT OPERATIONS

2-23. The purpose of MISO is to induce or reinforce foreign attitudes and behavior favorable to the originator's objectives. In simpler terms, MISO seek to change or reinforce foreign attitudes to further U.S. national objectives.

2-24. The more deliberate the planning process is prior to execution, the more likely the MISO effort will be coordinated and integrated with the supported unit's plans. Planning for MISO begins the process of identifying specific individuals, groups, or organizations to influence as part of the commander's overall objectives. When the MISO process is properly executed, it provides the commander with a formidable tool to gain a decisive advantage on the battlefield and potentially save lives.

2-25. Key terms discussed throughout this chapter are defined below:
- *MISO programs* support U.S. national policy and objectives and are approved by the Under Secretary of Defense for Policy (USD[P]) through the interagency process. Approved MISO programs provide the framework for the execution of MISO in support of the range of military operations. MISO programs include objectives, themes to stress, themes to avoid, potential target audiences, attribution posture, means of dissemination, a concept of operations, and funding sources. To execute MISO, U.S. policy requires a USD(P)-approved MISO program be in effect

for the operation in question as well as specified execution authorities in the form of an execution order, operation order, or theater security cooperation plan.
- A MISO *theme* is an overarching subject, topic, or idea. It often comes from policymakers who establish the parameters for conducting MISO by delineating the themes to stress and avoid.
- *Target audiences* are groups of people that can effect achievement of the commander's mission and toward which actions are directed. Planning for MISO requires a thorough analysis of each group's attitude, behavior, susceptibility, and sources of information to determine which themes, messages, and means will effectively influence the group to exhibit the desired behavior.
- *MISO objectives* are general statements of measurable response that reflect the desired behavioral change of foreign target audiences and best support the accomplishment of the supported commander's mission. Generally, the MISO objective is written at the geographic combatant commander level and is part of the geographic combatant commander's overall campaign plan.
- *Supporting MISO objectives* are the specific behavioral responses desired from the target audience to accomplish a given MISO objective. Supporting MISO objectives are unique for each MISO objective, and each MISO objective always has two or more supporting MISO objectives.
- *MISO series* consist of all the MISO products and actions designed to accomplish one behavioral change by a single target audience.
- *Distribution* is the movement of completed MISO products from the production source to the point of dissemination.
- *Dissemination* is the delivery of MISO products directly to the target audience. Planners must keep in mind that each target audience varies greatly in their access to a particular medium, whether it is radio, television (TV), newspapers, posters, and so on. Additionally, the ability of target audiences to understand the message varies because of language, cultural, or other barriers.

Note. An essential element of effectively planning for MISO is integration into the targeting process. This allows MIS forces the opportunity to get the rest of the staff to understand the importance of its nonlethal targets and the psychological effects of planned operations on target audiences.

2-26. Tactical-level MISO are typically conducted by MIS forces (directly attached at the SFODA level) through close-range means, such as face-to-face meetings, loudspeaker broadcasts, or by pinpoint distribution of products, such as leaflets to a particular village.

2-27. Strategic-level MISO focuses on conveying select information to international regional foreign audiences. Operational-level MISO focuses on a theater of operations, whereas tactical-level MISO focuses on conveying select information inside a tactical unit's AO. Because MISO are planned and executed at all levels, it is important that they be mutually supporting or complementary of other United States Government (USG) information activities, as well as other information capabilities (for example, PA and OPSEC). Generally, tactical-level MIS units will use this higher-level guidance as the basis for their own information activities.

Note. The proximity to the target audience does not determine the level of support (tactical, operational, or strategic). Mission analysis and, ultimately, the MISO objective determine the level of support. Likewise, the impact of a MISO effort at the tactical level can have operational or strategic implications.

2-28. Generally, a joint MISO task force assigned to the higher joint command provides direct guidance to all MIS forces in-theater to coordinate MISO at the strategic, operational, and tactical levels. Depending on the mission, either a company- or detachment-sized MIS element will typically support the JSOTF or a SOTF. Task organization is tailored in accordance with the mission, available resources, and priority of effort.

2-29. The primary mission of a MIS company supporting a JSOTF is to conduct operations that influence behavioral responses and advise the commander of those responses and their impact on the operation. The

MIS company typically supports a variety of tactical-level special operations forces missions, monitors the psychological state of target audiences in the operating environment, and analyzes adversary information activities. It can develop, produce, and disseminate tactical-level products within the guidance assigned by the approval authority. The company consists of a company HQ section with a span-of-control of three to five detachments. Development and production of MISO products are typically conducted at the company level.

2-30. A MIS detachment attached to the SOTF is comprised of a HQ section and three to six teams comprised of three to five Soldiers. The MIS detachment provides direction and oversight of the teams. The MIS detachment does product-dissemination planning by determining dissemination priorities and tracking the dissemination of products within the AO. (FM 3-53, *Military Information Support Operations,* and FM 3-05.301, *Psychological Operations Process, Tactics, Techniques, and Procedures,* provide a complete description of the responsibilities and duties of the respective MIS elements.) The team is a three- to five-man element led by an E-7; the team is generally task-organized down to the SFODA. The E-7 also serves as the MIS planner and advisor to the SFODA/SFODB commander and is responsible for the integration and employment of the team. The MIS team is the link between the SF commander and local target audiences in a given AO. This linkage is possible through face-to-face communication and rapport-building with local nationals. Whether conducting loudspeaker operations in support of combat operations or collecting information, on target audiences and the operational environment, the MIS team is a tactical asset that can significantly influence overall operations. Because of its integrated nature, the team is most effective when maintained as a cohesive element. After-action reviews suggest the division of a team decreases its capability.

2-31. A tactical MIS team can perform the following functions during combat operations:
- Reduce the adversary's will to fight. MIS Soldiers can use loudspeakers and leaflets to instill and exploit the fear of death or defeat in the adversary; undermine the adversary's confidence in their leadership; decrease their morale and combat efficiency; and encourage surrender, defection, or desertion.
- Support deception activities through employment of loudspeaker assets and other means.
- Minimize civilian interference with military operations.
- Monitor and assist in efforts to counter propaganda in the AO.
- Plan, develop, and monitor a key-leader engagement strategy for supported commanders to ensure this critical influence tool is appropriately aimed at achieving the commander's objectives.

2-32. During stability and support operations, in addition to discouraging civilian interference and assisting in efforts to counter propaganda, MISO can support the following:
- *Humanitarian assistance.* MIS units support humanitarian assistance operations by providing information on program benefits, shelter locations, food and water points, and medical-care locations. MIS units also publicize humanitarian assistance operations to build support for the United States and HN governments.
- *Peacekeeping.* MISO help gain acceptance for U.S. or allied forces in the AO, thereby gaining support and compliance with U.S. and allied policies and directives, and increasing support for HN governments or military and police forces.
- *Noncombatant evacuation operations.* MIS units support these operations by reducing interference from friendly, neutral, and hostile target audiences and by informing evacuees.
- *Demining operations.* MISO educate the target audience on the dangers of mines, how to recognize mines, and what to do when a mine is encountered. MIS units encourage target audiences to report locations of mines and unexploded ordnance.
- *Foreign internal defense.* MISO help build and maintain support for the HN government and its forces while decreasing support for insurgents.

2-33. In the execution of the MISO series, the team leader coordinates the dissemination of all products through the higher MIS detachment and the maneuver unit. Products normally comprised of standard visual products, such as posters, handbills, and novelty items; audio products, such as loudspeaker broadcasts, radio messages, and compact discs; and audiovisual products, such as video compact discs or digital videodiscs. In dissemination, the team follows the guidelines set forth in the series dissemination work sheet, which gives specific instructions for required dissemination to the appropriate target audiences. As part of its support to MISO development, the team provides details on possible dissemination sites and optimal dissemination times.

2-34. Whatever dissemination means are used, the MISO messages communicated to the target audiences are guided by themes. A theme is a subject, topic, or idea used as a planning tool to develop a MISO series. For MISO, a theme is developed through target audience analysis based on approved MISO objectives and formulated to affect the attitudes or behaviors of the target audience. As such, themes are broad, somewhat static, and not communicated to the target audience. At the tactical level, MISO themes should be the basis for all communication with the adversary, local populace, and any other target audiences. The two types of themes are the themes to be stressed and the themes to be avoided. The list of themes to stress and avoid can be found in the higher HQ MIS annex. The following are examples of themes:

- *Themes to stress.* Only local people can resolve problems, coalition forces do not favor any group or faction, and displaced persons should return to their homes. Themes to be stressed vary according to the target audience:
  - *Enemy forces.* Themes include inevitability of defeat, hardship and privation, and absence from loved ones.
  - *Local population.* Themes to stress include security and stability, reconstruction and economic prosperity, tribal and cultural, nationality and history (for example, Iraq is a multi-ethnic, tribal-sect-dominated state), insurgents are criminals and miscreants.
  - *Foreign governments.* Themes to stress include commitment and resolve, international security, and cooperation.
  - *Third-party organizations.* Themes to stress include security, stability, and solidarity with military forces.
- *Themes to avoid.* Themes to avoid include religious issues, cultural comparisons, women's roles in local society, themes that appear to favor one faction or group over another, and themes that degrade local ethnic, cultural, or religious values.

2-35. A message is a communication of the theme, whether visually, audibly, or in written form. Messages are communicated to the target audiences to influence their attitudes and behavior. As such, messages are specific, constantly evolving with the situation, and tailored to specific target audiences. Messages may take either of the following forms:

- *Spoken.* Messages can be spoken communication delivered in TV and radio broadcasts, talking points delivered during face-to-face communication or loudspeaker broadcasts.
- *Written.* Written messages can be delivered by leaflets, handbills, or posters.

In any case, a message is a single thought to be conveyed from U.S. forces to the enemy or other target audience (such as the local populace).

2-36. Because of the number of messages and themes available to the commander for his information activities, it is imperative that the IO staff coordinate and synchronize all messages (for example, MIS planner, PAO, CA planner) emanating from the commander. This deconfliction should ensure that messages from different elements are not contradicting one another and that the correct message is communicated to the correct target audience at the right time and place.

2-37. MISO are not the only IO capability that produces themes and communicates messages to the adversary or populace. PA produces and uses PA themes and messages to communicate with the media and inform the populace. MILDEC may communicate deceptive messages to the adversary. Therefore, MISO, PA, and MILDEC must coordinate and synchronize themes and messages so that the correct message is communicated to the correct target audience at the right time and place to avoid information fratricide.

2-38. By using MISO, the commander brings to bear a force-multiplier that uses its capabilities to degrade the enemy's will to fight, reduce civilian interference, minimize collateral damage, and maximize the local population's support for operations. MIS forces do this by using assets at their disposal to reach local and over-the-horizon targets with different visual, audio, and audiovisual products. Proper employment of

MISO means fewer casualties (friendly, enemy, and civilian), fewer U.S. troops required to accomplish the mission, shorter operations, and less damage to infrastructure. Success in these areas results in faster reconsolidation of combat elements, less required rebuilding of infrastructure, and ultimately, quicker return of a nation to self-sufficiency.

2-39. When employing MISO, there are three primary limitations to consider. First, only personnel school trained in MISO and designated by their Service as MIS officers or Soldiers should develop MISO series (although any friendly-force element can disseminate products and conduct face-to-face engagements). Second, there are legal and political factors that may restrict the use of MISO. MISO must follow U.S. and international laws, especially when used without a declaration of war. The third constraint is time. MISO planning must begin early in the operation and continue throughout if it is going to effectively influence the target audience in time to support the operation. FM 3-53 and FM 3-05.301 provide further details on MISO planning and TTP.

ELECTRONIC WARFARE

2-40. EW plays a major role in attacking and exploiting the adversary's ability to use information, while defending the U.S. ability to process information. By definition, EW is any military action involving the use of electromagnetic and directed energy to control the EMS or to attack the enemy. It consists of three divisions: electronic attack (EA), electronic protection, and electronic warfare support. EA is the use of electromagnetic energy, directed energy, or antiradiation weapons to attack personnel, facilities, or equipment with the intent of degrading, neutralizing, or destroying enemy combat capability. EA is considered a form of fires. Electronic warfare support involves actions taken to search for, intercept, identify, and locate or localize sources of intentional and unintentional radiated electromagnetic energy for the purpose of immediate threat recognition, targeting, and planning. Electronic protection involves passive and active means taken to protect personnel, facilities, and equipment from any effects of friendly or enemy employment of EW. Of EW's three major components, EA has the most direct role in support of IO, especially at the tactical level. At the tactical level, EA is primarily used to attack adversaries by jamming the electromagnetic frequencies used by the adversary (degrading or disrupting information flow) or using the EMS to deceive the adversary (affecting the information content available to decisionmakers).

2-41. The objective of EA planning and execution is to ensure use of the EMS for friendly forces while preventing the adversary's effective use of the spectrum. Electromagnetic jamming denies the adversary the use of its receivers by overwhelming them with high-powered signals on the same frequency as the receiver. Procedurally, this is accomplished through spectrum management and deconfliction. Spectrum management controls frequencies that all friendly emitters use to prevent interference and fratricide. Deconfliction is the process used to avoid conflicts in frequency usage. It is also used during mission execution to resolve interference problems.

2-42. To meet the technical requirements of EA, an EW officer is often needed to effectively plan, coordinate, monitor, and assess friendly and adversary force activity in the EMS. However, depending upon the unit, an EW officer may not be authorized. If the unit does not have an EW officer, EA responsibilities will usually fall to the IO planners. In such a case, IO planners should research the EW assets available to the command from higher HQ and establish contact with representatives from the units that possess the assets.

2-43. Synchronization of EW tasks is imperative. Up to a certain point in the operation, friendly forces may want adversary decisionmakers to communicate, so they do not want to tip off the adversary by jamming too early in the operation. Additionally, lack of frequency deconfliction can result in information fratricide among friendly forces.

2-44. Basic planning considerations for EA include the following:
- *Enemy vulnerabilities.* The J-2/G-2/S-2 should have an electronic order of battle and other intelligence products that identify how the enemy uses the EMS—which systems are critical to adversary operations, what those systems are used for, and what frequencies those systems use. This information helps determine specific electromagnetic vulnerabilities that may be exploited.

- *Friendly capabilities.* Tactical units have limited organic EA assets. Most EA assets are assigned at higher echelons. It is important to know specific capabilities organically available to the unit, as well as those assets that are available through higher HQ.
- *EA deconfliction.* Frequency-spectrum management ensures effective use of the EMS, preventing interference with military and civilian frequencies.
- *Rules of engagement.* EA usually has rules of engagement that must be followed to avoid violating international treaties that control the use of the EMS. In peacetime, EA is generally used only to protect friendly forces. In wartime or conflict, there are restrictions concerning the impact of EA on civilian emergency services.

Note. At the SFODA/SFODB level, leaders must coordinate with the SOTF IO planner for an effect in the EMS. The IO planner will coordinate with the EW officer to identify the means to achieve the effect. It is imperative that EA be coordinated and deconflicted with higher and adjacent elements. The frequency range on the EMS or the cell tower that needs to be jammed for an operation may be the primary means of communications or intelligence collection for friendly forces.

COMPUTER NETWORK OPERATIONS

2-45. CNO provide IO planners with the ability to affect information content and flow within cyberspace. The three elements of CNO are computer network attack, computer network defense (CND), and computer network exploitation. A computer network attack is the use of computer networks to disrupt, deny, degrade, or destroy information resident in computers and computer networks, or the computers and networks themselves. CND is the use of computer networks to protect, monitor, analyze, detect, and respond to unauthorized activity within Department of Defense information systems and computer networks. Information technology professionals assigned to the command, control, communications, and computer systems directorate of a joint staff (J-6); assistant chief of staff, command, control, communications, and computer systems staff section (G-6); command, control, communications, and computer systems staff officer (S-6); and specialized organizations typically execute CMD. Often these same personnel conduct IA activities. Computer network exploitation is an enabling operation and intelligence collection capability conducted by using computer networks to gather data from target or adversary automated information systems or networks. IO planners strive to integrate these CNO elements in cyberspace while synchronizing and coordinating them with other IO capabilities to ultimately achieve information superiority. CNO is typically a collective effort involving separate Services and interagency organizations seeking to achieve effects across the globally interdependent network of information technology infrastructures that include the Internet, telecommunications networks, computer systems, and embedded processors and controllers.

Note. Other than CND, which is executed by the J-6/G-6/S-6, CNO is not conducted at the JSOTF. If a need for CNO arises, the IO planner at the JSOTF will request support through appropriate channels. It requires a long lead time for the approval process.

2-46. Although CNO has a unique request process involving the joint operations planning process and classified supplemental processes, successful CNO planning relies on a five-step process (Figure 2-7, page 2-17) that should be integrated into the unit's planning process.

1. **Identify CNO contributions to the information operation and determine targets.** IO planners should refrain from identifying specific tasks to CNO and should focus on the desired effect for CNO in support of the information operation. This can be achieved by reviewing the previously determined IO objectives and identifying which objectives can be supported by CNO in terms of the effect included in the objective. Once the objectives to be supported are identified, IO planners coordinate with J-2/G-2/S-2 to identify potential targets for CNO. Related planning factors for a typical computer network attack include the—
    - Quality of available intelligence.
    - Reliability of access to the targeted network.
    - Level of risk determined through a collateral-effects estimate.
    - Timely availability of a required capability.

    IO planners should include CNO-related activities when developing measures of effectiveness (MOEs) associated with the IO objectives and identify collection methods for MOEs associated with CNO.

2. **Determine time available.** During this step, IO planners determine if there is time available to employ CNO in support of the information operation. Considerations include approval timelines, required timing and execution of other IO capabilities in regard to shaping the information environment, and whether delayed or disapproved CNO integration will substantially affect specific IO capability efforts or IO's overall contribution to the mission.

3. **Obtain approval.** Although CNO is an operational-level planning function, units at the tactical level should not hesitate to request CNO support if they identify an appropriate use. The request for CNO is a request for national-level assets. The CNO approval process has unique complexities. CNO-related authorities are retained at high echelons, and IO planners should plan accordingly. When completing CNO requests, IO planners should clearly show the necessity for CNO and its impact on mission accomplishment. IO planners at the tactical level should contact operational-level IO planners to submit CNO requests. Operational-level planners should handle the somewhat-formal approval process. One significant factor pertaining to approval is the fact that computer network exploitation has two distinct subelements. The first is that of an intelligence function (collection) and the second is that of an operations activity (enabling). This distinction is important because CNO, by doctrinal definition, does not include the intelligence subelement of computer network exploitation. However, under Title 10, United States Code, CNO can conduct similar activities as long as they are not for the purposes of intelligence collection. The intelligence community, operating under Title 50, United States Code, can conduct both subelements of computer network exploitation. Therefore, to ensure all activities are legal, coordination between these communities is usually required prior to approval.

4. **Execute CNO tasks.** Execution of CNO should be monitored by the unit to ensure synchronization with other IO capabilities as well as unit maneuver elements. IO planners should identify collected CNO-associated MOEs and make recommendations to the commander concerning variations to the plan.

5. **Conduct after-action review.** Upon completion of the operation, IO planners should determine if CNO tasks supported the IO objectives, as planned. IO planners should identify additional CNO tasks that may have increased friendly-force advantage. IO planners also review the request for CNO and determine what information would have expedited the request. They determine if all MOEs were able to be collected. If not, they determine what changes should be made for future operations? Finally, IO planners determine how the unit will handle unexpected CNO-related adversary responses in the future.

Figure 2-7. Five-step computer network operations planning process

COMBAT CAMERA

2-47. JP 1-02 defines COMCAM as *the acquisition and utilization of still and motion imagery in support of operational and planning requirements across the range of military operations and during joint exercises.*

Chapter 2

> **No, We Did Not**
>
> In the spring of 2009, Iraqi Security Forces conducted an operation in central Iraq with U.S. SF advisors to capture an insurgent leader. During the course of the operation, sensitive-site exploitation information led the FID partner-force to a second location. The Iraqi Security Force with U.S. SF advisors searched the second location and then departed. Following the operation, the government of Iraq received complaints that property had been damaged and personnel were harmed during the search.
>
> Due to the IO planning for the operation, COMCAM accompanied the assault force and provided crucial footage proving that no damage was caused to the facility and personnel were treated with respect. The footage was reviewed and released by higher HQ to refute the accusations.

2-48. COMCAM documents military operations with both video and still photography. When the mission dictates, it is used to—

- *Gather intelligence.* COMCAM provides imagery of potential targets or target areas and supports battle-damage assessments.
- *Support planning efforts.* COMCAM validates assumptions by providing accurate images of a situation.
- *Provide imagery to PA and other IO capabilities.* COMCAM provides graphics, photography, video products, and print media to enhance the effectiveness of PA press releases and MISO products.
  Document interrogations and autopsies. COMCAM provides evidence of proper techniques and procedures.
- *Support landing zone studies.* COMCAM imagery can help determine the diameter of the area and the terrain's grade.
- *Provide historical documentation.* COMCAM provides evidence of events for future use (for example, Red Cross investigations) and preserves the accuracy of historically significant events.

2-49. COMCAM captures a photographic record of military operations but, more importantly, it allows commanders to provide visual proof of operations for MISO and PA and to counter adversary IO for enemy, adversary, and neutral information activities. Once the COMCAM team's captured imagery or video is released by the supported commander, the imagery is digitally transmitted to the noncommissioned officer in charge of the JSOTF. Prior to sending the product to the joint combat camera center (JCCC), the imagery is reviewed to obtain the commander's release approval. Once approved, the image is sent to the JCCC, where it may be used by any Department of Defense entity that has access to the imagery database. For example, MIS elements may use these pictures to develop products, whereas PA will use these pictures during press conferences and media-engagement activities. At the tactical level, units can use these images to make immediate impact on the populace within the operating area by producing visual products highlighting local events, good or bad, to achieve the goals set by the unit commander. The complete cycle from image acquisition to receipt by the JCCC must occur within 24 hours for the collection imagery to remain a viable decisionmaking tool for national-level leaders.

2-50. The method of documentation depends on the purpose of the mission, the environment in which the documentation occurs, and the support available to the Soldiers documenting the event. There are three different COMCAM documentation methods:

- *Still photography.* Still photography involves producing, processing, and reproducing still-picture films, prints, and transparencies. These images can be captured using film or digital cameras or can be taken from motion picture or video photography. COMCAM teams use digital still-video cameras to capture and transmit images electronically. Some cameras also have night-vision devices that permit them to be used during darkness or other limited-light conditions.

- *Motion media.* Motion media is documentation of activities or operations as they occur. Motion media technology can be used in daytime, nighttime, and limited-visibility operations. The film, which captures positive and negative images, must be edited before IO or other staff elements can effectively use it.
- *Multimedia presentations.* Multimedia products can be used for a variety of purposes, from meeting training requirements to serving as a means to transmit public information. They allow commanders to review the operations and training of their forces, and introduce new and improved operational techniques and developments to subordinates.

2-51. COMCAM imagery must be reviewed by appropriate staff members of the supported command prior to release. The supported commander is the release authority for all collected COMCAM images before they are transmitted out of theater. Composition of the review board should be tailored based on the specific unit design. A typical review board includes the following individuals:

- J-2/G-2/S-2 representative (for identification of possible intelligence and exposure of classified information).
- OPSEC officer (for identification of possible disclosure of unit critical information [EEFI]).
- Judge Advocate General (for identification of possible or perceived violations of the laws of land warfare).
- Operations directorate of a joint staff (J-3)/assistant chief of staff, operations staff section (G-3)/operations staff officer (S-3) representative (for identification of exposed TTP or any content that is not desirable for release).
- PAO (for public-release consideration).

2-52. To maximize COMCAM support, leaders should—

- Employ COMCAM as an operational asset assigned to the J-3/G-3/S-3. A COMCAM representative is identified within the J-3/G-3/S-3 to plan for the employment of COMCAM.
- Plan to employ COMCAM during the initial phases of an operation to ensure comprehensive mission documentation.
- Provide COMCAM with full mission access (as is reasonably and tactically feasible) during each phase of the operation.
- Ensure COMCAM coverage availability before, during, and after operations.
- Ensure tasks to COMCAM personnel include clearly defined requirements and priorities. Include a purpose for each task to take advantage of COMCAM personnel initiative.
- Ensure COMCAM imagery is reviewed by PAO prior to release outside of the organization.
- Ensure COMCAM personnel provide imagery to the JCCC for immediate distribution to support strategic and operational objectives.

Note. COMCAM can provide images for MISO, MILDEC, PA, and CMO. It can be used for battle damage MOEs. It can also serve as a record.

LOCAL POPULACE AND KEY-LEADER ENGAGEMENTS

The people...represent many things in [a] conflict—an audience, an actor, and a source of leverage—but above all, they are the objective. The population can also be a source of strength and intelligence and provide resistance to the insurgency. Alternatively, they can often change sides and provide tacit or real support to the insurgents. Communities make deliberate choices to resist, support, or allow insurgent influence.

General Stanley A. McChrystal

2-53. Engagement of the local populace and key leaders is an important part of any counterinsurgency campaign or operation. Effective communication with key leaders and key communicators can be critical to mission success at all levels. Not only do commanders and senior leaders conduct key-leader

Chapter 2

engagements, but every Soldier has the potential to conduct some form of engagement with the local populace and leaders, and most importantly, to communicate a message through their actions.

2-54. Military leaders who prepare, listen well, and communicate effectively are more likely to gain the cooperation and trust of the local populace. Commanders must also understand that influencing a given populace will most certainly require committing assets to help meet the people's needs—talk will only go so far. People associate actions with messages. SFODA actions on the objective must support the command's themes and messages. Messages with few or no supporting actions generally are given little credibility.

2-55. The attached MIS element is the most capable, by purpose, training, and organization, to develop, plan, monitor, and assess a commander's key-leader engagement strategy. As key-leader engagement is targeted to influence an action or nonaction, it falls to the MIS element to develop the appropriate messages to be disseminated to the appropriate target audience at the right time as part of the commander's larger influence efforts. By using attached MISO personnel to manage the KLE program, a commander ensures a broader, more effective influence effort in his AO.

TYPES OF ENGAGEMENTS

2-56. In interacting with the local population, there are well-planned and coordinated meetings and chance encounters, as described below:
- *Planned face-to-face meetings.* These meetings are daily or weekly key tactical-leadership activities that occur with local leaders and populace. Planned face-to-face engagements are well thought out and resourced. They are an important facet of the mission during counterinsurgency. Face-to-face meetings are often the result of the targeting process and support specific effects tied to accomplishing the desired end state.
- *Key-leader engagements.* These meetings are at all levels between military leaders, HN government, and tribal or village leaders to achieve or support a specific desired effect. To match the appropriate military leader with the leader being engaged requires careful planning.
- *Chance encounters and contacts.* Chance encounters typically occur with the local populace during patrols and other mission activities by MIS teams, Civil Affairs teams, and SFODAs. All personnel with access to the local populace and leadership should be briefed on how to conduct face-to-face engagements, be aware of current matters of interest to the local populace, and be knowledgeable of the command's themes and messages.

THEMES AND MESSAGES

2-57. Themes and messages are two distinct entities. Each has its own purpose—they are not interchangeable. Themes are usually associated with specific lines of operations and are planning tools that guide the development of messages and other information tools (for example, talking points, MISO print and broadcast products, and PA guidance). Themes represent the broad idea the commander wants to get into the mind of the target audience. Themes are not communicated to the target audience; that is the role of messages. Themes are broad and enduring.

2-58. Messages support themes and are communicated by speech, writing, or signals. They contain the information that will be delivered to the target audience. Messages are tailored to specific audiences and are meant to elicit or prevent a certain behavior. Messages constantly change with the situation and mission. Sources of messages include the following:
- *Command information messages.* These messages convey the policies and intent of local commanders to their subordinates. The PAO develops command information messages.
- *Public information messages.* These messages convey information to local target audiences through news, public-service information, and announcements from HN officials. The PAO develops public information messages.
- *MISO messages.* These messages convey specific information to selected foreign audiences to influence their attitudes, perceptions, beliefs, and behavior. MIS elements develop these messages.

Types of Messages

2-59. Messages may be either negative or positive. Negative messages are used to attack the target audience or to convey the likelihood of negative consequences if the target audience does not engage in the desired behavior (for example, if you do not surrender, you will be killed). Positive messaging offers specific or implied benefits if the target audience engages in a certain behavior (for example, cooperation with coalition forces will result in the construction of a health clinic).

> *Note.* By doctrine, there are no IO themes and messages. MISO and PA have themes and messages. If needed, in coordination with MISO and PA representatives, the IO staff may have to develop command themes and messages.

Message Development

2-60. When developing messages, it is important that message content addresses target-audience vulnerabilities (or perhaps an interest or motivation). Target-audience vulnerabilities are determined by considering the following four factors:
- *Motives.* Look for factors that drive target-audience behavior. Primary motives include basic life needs such as shelter, security, and food. Secondary motives evolve from social interaction within the family, clan, or tribe, or from membership in political and religious organizations.
- *Demographics.* Look for target-audience characteristics, such as gender, ethnicity, religion, and age. Planners must determine which characteristics can be exploited to affect target-audience behavior.
- *Psychographics.* Look for the target audience's cognitive characteristics relevant to the world around them, both near and far. These can be values, beliefs, attitudes, and ideology that trigger emotional responses.
- *Symbols.* Symbols are a sophisticated mix of graphics, video, audio, or audiovisual objects that reference architecture, religious symbols, historical events, and symbols with cultural or contextual significance to the target audience.

2-61. Once vulnerabilities are identified, messages are crafted that communicate the approved themes to the target audience and address its vulnerabilities. Crafting messages is an art that requires time and thought. Figure 2-8, page 2-22, provides a sample message. A few guidelines to follow include—
- Limit each message to one thought.
- Keep each message succinct. Complex messages pose challenges for senders, translators, and receivers. Limit each message to one sentence and minimize internal sentence punctuation.
- Keep messages to a manageable number. Rule of thumb is no more than five messages per theme or target audience.
- Tailor messages for the means and method of delivery and the target audience.
- Convey a story (the theme) by arranging the messages from first to last. The sum of the messages should then tell the story (or theme).
- Place the bottom line up front and summarize at the end. The first message should contain the most important thought. The last message should restate the first message.
- Consider developing "escape" messages that leaders and Soldiers can use to deflect conversations away from the themes to avoid.

Chapter 2

Themes	Messages
It is inevitable that the insurgents will be defeated.	• While your leaders sleep safe in their warm beds, you are left to suffer in the cold and wind. • Your mothers will mourn the deaths of their sons, and your children will be orphans when you meet the bloody death that awaits you. • Lay down your weapons and return home to the families who need you.
The Army is honorable and capable.	• The Army is the guardian of the people. • The Soldiers fight like bold lions for the freedom of the nation. • The enemy comes with foreigners in the night to murder and rob their fellow tribesmen. • Help the Army defeat its enemies and provide information about terrorists, weapons, people, and activities.
The insurgents are responsible for civilian deaths.	• The United States and its allies do everything possible to avoid civilian deaths. • The insurgents hide among the populace. • It is well known that the terrorists place women and children in harm's way when it suits their purposes.

Figure 2-8. Example message (paired to themes)

2-62. The tool that can be used to develop and organize themes and messages is a message development matrix (Figure 2-9).

Target Audience	Target Audience Vulnerability	Desired Target Audience Action(s)	Themes	Messages
People in Village X.	Security from villagers in Town Y.	Halt violent demonstrations.	Violence does not solve any problems.	• Violence does not improve your situation. • Further violence will lead to the withdrawal of coalition aid and support.

Figure 2-9. Example message development matrix

PREPARING FOR A FACE-TO-FACE MEETING

2-63. Conducting planned and unplanned engagements with the local populace and their leaders requires preparation to be effective. Time spent researching the target, anticipating requests and issues, and rehearsing the meeting often pays high dividends. Units conducting operations among the local populace should assume engagements will occur and prepare for them prior to the mission. Lack of preparation may lead to embarrassing situations that have the potential to diminish the effectiveness of friendly forces and create an advantage for the adversary.

2-64. Leaders should consider the following, when preparing for a key-leader engagement or planned meeting:

- *Identify upcoming meeting.* Meetings may occur because of—
    - *A recurring schedule.* Unit leaders often have schedules for reoccurring meetings, such as weekly city council meetings or occasional sit-downs with elected officials, clan or familial

leaders, or tribal sheikhs. These meetings present a preplanned opportunity to conduct a key-leader engagement.
- *Direction from higher HQ.* A higher HQ may direct a unit to engage a specific leader or group to support its objectives. Units should incorporate higher HQ directions into the existing schedule of meetings.
- *Mission planning.* During mission planning, key-leader engagements may be identified as part of the operation. These engagements may have to be conducted outside the normal schedule of meetings.
- *Requests from local leaders.* Local leaders sometimes seek a meeting to address specific concerns or emergencies. When arranging these meetings, always consider the importance of pairing the right military leader with the right civilian leader. Leaders should be very cautious about giving access to relatively insignificant or noninfluential civilian leaders who may try to gain access to senior military leaders.

Note. After identifying a meeting, leaders should determine an appropriate location for the meeting. If hosting the event, leaders ensure the area is presentable and cleared of any operational information. They ensure the location is quiet and away from disturbances such as phones or radios.

- *Identify target-audience characteristics.* Gather as much information about the local leader as possible (for example, proper name and title, approximate age, family members, ethnicity, language spoken, and relationships to other leaders, friendly forces, third-party organizations, and the adversary). One way to obtain this information and prepare for a meeting is to consult personnel who have met with the person before. It is important to continually refine and update background information based on experience with the individual. Characteristics of the local leader that are good to know include—
  - *Language spoken.* Identifying the leader's language may be difficult if multiple languages and dialects are spoken in the operational area, all of which could require different interpreters.
  - *Education/literacy level.* Level of education may determine which form of the language the local leader speaks. Many languages have a colloquial version and a more formal textbook version (often referred to as high or formal). Knowing the individual's level of literacy may impact on decisions to leave written products during the meeting.
  - *Customs and etiquette.* Identifying unique customs and proper etiquette prevents awkward moments during the meeting.
  - *Attitudes.* Understanding the local leader's attitudes toward military forces and toward other organizations and groups in the AO helps to avoid a tense discussion. Plan a strategy to overcome the leader's negative perceptions.
  - *Key advisors.* Knowing the local leader's key advisors facilitates follow-on discussions and may bridge gaps that cannot be resolved with the target directly. Advisors are also helpful in gauging the leader's perceptions of U.S. forces and of the results of the meeting.
- *Identify target-audience concerns.* Every individual engaged has some key concerns that may be raised during the meeting. Being prepared to address these concerns will greatly facilitate communication. Some of the local leader's concerns may include—
  - *Local conditions.* Religious, public health, crime, and economic issues in the key leader's AO may impact the dialogue during the engagement.
  - *Needs.* The local leader will probably discuss the needs of his followers. Needs can be the basic requirements of food, water, and shelter, or they can be more complex—political power-sharing, contact between the populace and U.S. forces, or getting help with reconstruction or security matters. Anticipating the leader's concerns allows the U.S. leader to plan resources that can enhance cooperation.

Chapter 2

- *Religious, political, and economic viewpoints.* Conditions in the operating area may shape religious, political, and economic viewpoints, but viewpoints vary depending on the individual. Knowing the local leader's viewpoints makes it easier to address or avoid sensitive topics that may detract from the meeting. This does not mean that these topics should be avoided, but in general it is best to avoid sensitive topics until one is more familiar with the target audience.
- *Review previous meetings.* Notes, comments, and debriefings from previous meetings with the key leader may indentify previous agreements made with the local leader and reoccurring topics that are likely to be raised at the meeting. For example, if during previous meetings with a city mayor, the mayor asked for more money at each meeting, chances are that he will ask for money again.
- *Identify end state.* Typically, the engagement's end state is the action (or, at times, the inaction) friendly forces want the target to take. For example, the unit may want a leader to actively support Army recruiting or a religious leader to stop encouraging violence.
- *Develop messages.* Develop tailored messages that support the engagement's end state and address the key leader's vulnerabilities. When possible, use already approved messages. Review MISO themes to stress and avoid.
- *War-game responses and reactions.* Develop appropriate counteractions to the key leader's most likely responses and possible demands.
- *Develop meeting exit strategy.* Have an exit strategy so the meeting can end tactfully.

2-65. Leaders can use a face-to-face engagement work sheet (Figure 2-10, page 2-25) to plan critical aspects of key-leader engagements. When kept current, the work sheets are a useful planning tool for future engagements with the same target. Commanders and leaders should also rely heavily on assigned or appropriate themes and messages for reinforcement during the engagement. MIS forces habitually do target audience analyses on various individuals, groups, and factions within their assigned AO and are skilled at planning operations to influence.

CONDUCTING A FACE-TO-FACE ENGAGEMENT

2-66. The following guidelines can help ensure a productive engagement. The spokesman should—
- Position himself immediately next to the engaged key leader and designate a second person to be a recorder.
- Establish rapport with the target audience. The spokesman uses a greeting phrase in the native language, when possible. The spokesman arranges for seating and offers something to drink for the meeting.
- Introduce everyone in the party and record the names and positions of everyone in attendance.
- Avoid rushing through the meeting. The spokesman plans for enough time to accommodate the culture and avoids making the target audience feel they are low on the priority list. The spokesman is prepared for small talk before discussing business. He takes cues from the target audience.
- Ask permission to take photos of the target audience.
- Apologize in advance for any cultural mistakes made. The spokesman assures the target audience that he does not mean to offend and asks that the target audience identify any mistakes made. The spokesman is careful about telling jokes; they can backfire when translated.
- Avoid restricted topics and confrontational attitudes.
- Never assume that the target audience does not speak or understand English.
- Always maintain eye contact with the person he is speaking with, not the translator. The translator is his voice. The spokesman communicates through the translator, not to him. He watches the target audience's gestures, eyes, and body language, not those of the translator.
- Speak in short clips. He should not recite a long paragraph and expect the translator to accurately convey the message. The target audience should feel like he is being conversed with, not being lectured to. He should remember that one to two sentences at a time is a good rule.
- Avoid using acronyms, slang, and idioms. He should keep the language simple.

- Treat all members of the target audience with courtesy and respect.
- Avoid making or implying promises that cannot be kept.
- Avoid elevating his position or embellishing his authority. Although he certainly may have to check with higher authorities before making promises or decisions, using it as an excuse too often may decrease the target audience's respect for him as a leader.
- Use open-ended questions to facilitate discussion. Yes or no answers tend to be incomplete and inaccurate.
- Be aware of the body language from all parties. He ensures that the body language does not negate the message.
- Recap what has been said, as the meeting closes, and clarify expected actions by both parties.

Target:	Date-Time Group:	Location:
Intended target is John Smith.	210900DEC11	FOB Bragg, Building 2, Room 123
Characteristics:		
Records from previous meetings indicate John Smith is a stern tribal leader. Research shows him to be supportive of U.S. interests, but he has been known to support operations against U.S. forces when it benefits him personally. A number of Smith's acquaintances verify this information. Meeting will be limited to one hour.		
Environment and Concerns:	**Previous Meetings:**	
Meeting at FOB Bragg is a friendly environment. Our goal is to provide assurance to Smith that the United States will support him upon his return to the tribal region.	Previous meetings have been cordial and show Smith appears to support U.S. interests. Smith has indicated a desire to partner with U.S. forces in the tribal region.	
Desired End State:	**Themes/Messages:**	
Reconfirm to Smith that the United States will continue to support his tribe by providing a detachment to support training against insurgency.	The message Smith must receive is that the United States will continue to provide support for training and combat operations. Posters and pamphlets clarifying this support will be provided to Smith upon his departure.	
Anticipated Reaction/Issues:	**Response:**	
Smith's reaction should be positive provided he is affirmed of continued U.S. support for his tribe. As an issue, it is anticipated that Smith will request monetary compensation for the families of wounded and killed tribal members.	Previous meetings indicate that Smith is a hard negotiator, and will initially request more than the United States is prepared to offer. Negotiators must remain steadfast in limiting promised support to that which is deemed appropriate.	
Meeting Strategy:	**Exit Strategy:**	
Meeting rehearsal is scheduled for 200900DEC11. The team leader and team sergeant will negotiate on behalf of the United States. Participants should be firm, but respectful. Negotiation will occur with Smith only. Samples of products will be provided. Promises of support must remain in keeping with commander's guidance.	Time limit for the meeting is set at 1 hour. The code word DISCONTENT will be used to end the meeting.	
Attendees:		
Scheduled attendees include John Smith, his aide (Michael Jones), the team leader, team sergeant, and intelligence sergeant. The intelligence sergeant will maintain a list of additional attendees, including approximate age, home town, contact information, profession, and demeanor.		
Notes:		
Team will conduct after-action review immediately following the meeting to compare notes and ensure an accurate understanding. Final report will be submitted to Colonel Jackson not later than 220900DEC11.		
Follow-up Actions:	**Next Meeting:**	
Meeting notes and after-action report will be provided to all detachments in the area. Any required coordination with higher will occur within 10 days of the conclusion of the meeting.	Next meeting is scheduled for 200900JAN12.	

Figure 2-10. Example of a face-to-face engagement work sheet

Chance Encounters and Contacts

2-67. During a chance encounter or contact with the target audience, the leader of the unit should conduct the face-to-face engagement based upon a preplanned battle drill, to include:
- *Security.* Protect both friendly troops and the target audience.
- *Time.* Limit the length of the engagement. Establish a codeword for when it is time to end the meeting.
- *Identify the local leader.* Ask who is in charge and talk to him. Otherwise, select a maximum one or two people to talk with. Do not distribute anything to the populace without the local leader's permission.
- *Take notes.* Get names of all people contacted, approximate ages, hometowns, businesses or activities, subjects covered, demeanor toward friendly forces, and any particular concerns of the target audience.
- *Establish rapport.* Offer the target audience refreshment (such as a bottle of water) and move to a comfortable location. Sit if possible.
- *Focus.* Stay on message by communicating the command's messages.
- *Report.* Report contacts with local leaders up the chain of command to ensure that an accurate picture of the situation is developed.

Working with Translators

2-68. Translators should be treated as a part of the unit. The better the translator is integrated into the unit, the better the translator's performance. Leaders must ensure translators are used for translation duties only. Using them for other activities may violate their contract. An example of misemployment is using a translator to run errands in town. However, sending the translator to town to coordinate a meeting for a U.S. official is allowed. A good rule of thumb is if the translator is acting as the leader's official voice, the action is legal.

2-69. Leaders must know each translator's strengths and weaknesses. The lives of Soldiers may be in the translator's hands. Translators should speak in first-person, remain nearby during engagements with the populace or key leaders, carry a notepad and take notes, project clearly, and mirror the leader's vocal stresses and overall tone.

2-70. The translator should be allowed rest periods to collect his thoughts. Meal meetings are especially challenging for a translator. Leaders should allow the translator to eat during or after the meeting.

2-71. Leaders must rehearse with the translator. If a translator performs poorly, it affects the target audience's perceptions of friendly forces. Rehearsals verify the translator's abilities, help identify words the translator may not know, and ensure the translator understands the overall message to be conveyed. This is especially important with complex, new, or sensitive issues.

2-72. The translator should be briefed on expected behavior. Leaders must recognize that translators are often seen as a representative of the command. All aspects of translator behavior must be kept professional and ethical, regardless of nationality or ethnicity. If operational details are briefed to the translator during the mission rehearsal, leaders should consider having the translator remain on the base camp until execution. Also, the translator should not have a cellular telephone or other communication device.

2-73. When using a translator, a leader must always maintain eye contact with his counterpart and not the translator. The leader communicates *through* the translator—not *to* the translator. The target audience should be observed for changes in gestures, postures, and body language. Leaders should speak in short clips—it is difficult to recite a long paragraph and expect the translator to accurately convey the intent.

COUNTERING ADVERSARY INFORMATION ACTIVITIES

2-74. Countering adversary information consists of programs of products and actions designed to nullify information for effect, misinformation, disinformation, and propaganda or to mitigate the effects of the information. Successful operations to counter adversary information require the use of all IO and other capabilities.

2-75. All elements of IO can and will support the operations to counter adversary information plans, but the focal point for such operations should remain with PA forces. *Adversary information* is used to describe information and activities used by an adversary or enemy, in peacetime and wartime, to undermine the legitimacy of operations and the credibility of the force. Previously the term *propaganda* was used to describe all forms of adversary information. The evolution of media forms and capabilities has made the term propaganda too limiting in describing how information is used by adversary states and nonstate actors to gain an advantage in the global information environment. To better clarify the use and application of adversary information this manual divides adversary information into the following four categories:

- *Information for effect.* Information for effect involves the use, publication, or broadcast of factual information to negatively affect perceptions and/or damage credibility and capability of the targeted group. Examples of uses of information for effect involve the premature announcement of collateral damage caused by friendly forces, reporting or images of the results of insurgent attacks on friendly forces, or release of captured sensitive, or classified information.
- *Propaganda.* Any form of adversary communication, especially of a biased or misleading nature, designed to influence the opinions, emotions, attitudes, or behavior of any group in order to benefit the sponsor, either directly or indirectly (JP 1-02).
- *Misinformation.* Incorrect information from any source that is released for unknown reasons or to solicit a response or interest from a nonpolitical or nonmilitary target (FM 3-13).
- *Disinformation.* Information disseminated primarily by intelligence organizations or other covert agencies designed to distort information or deceive or influence U.S. decisionmaker, U.S. forces, coalition allies, key actors, or individuals via indirect or unconventional means (FM 3-13).

Countering Adversary Information in Iraq

In the early part of 2009, U.S. forces were beginning to transfer bases to Iraqi Security Force control. During the planning of the transfer, a vulnerability was identified that enabled insurgent groups to exploit information and claim they had driven U.S. forces out of the bases. In the north, the Islamic State of Iraq propaganda efforts were active and could exploit upcoming transfers.

The SFODA identified Islamic State of Iraq activity in the vicinity of a future base transfer. The SFODA, in concert with the SOTF, developed a plan with conventional forces to counter Islamic State of Iraq propaganda and highlight the upcoming transfer to Iraqi Security Forces using multiple information capabilities, to include key-leader engagement, PA, PSYOP, and Iraqi Security Force engagement of media. This aggressive information operation with significant Psychological Operations support informed the population of the purpose for the base transfer, countered Islamic State of Iraq propaganda, and discredited Islamic State of Iraq in the area.

2-76. For the purposes of IO, adversary information justifies actions and bolsters legitimacy of an adversary. By communicating with the populace and, at times, friendly forces, the adversary offers a window into its philosophy, goals, objectives, and operations. Therefore, adversary information may provide a useful insight into how to defeat the adversary. Some of the more commonly used techniques include the following:

- *Name-calling.* This describes the use of a name or word to connect a person to something negative (for example, Muslim extremists describing or terming Westerners as Crusaders).
- *Glittering generalities.* This describes the twisting of the meaning of a word that has great symbolic value (for example, terming terrorist attacks as a jihad).
- *Euphemisms.* This describes the use of a milder word to make a situation seem less threatening (for example, "revenue enhancement" to describe a tax hike).

- *Transfer.* This describes the use of symbols to associate an agenda with a respected institution (for example, placing official letterhead on a piece of disinformation).
- *Testimonial.* Testimonials add credibility to a position (for example, using celebrities to testify on political issues).
- *Bandwagon.* This describes a technique which plays on the desire of people to fit in (for example, 7 of 10 workers prefer candidate X).
- *Fear.* This describes the manipulating of people's fears to elicit a behavior (for example, without jihad, the crusaders will invade your homes).

2-77. To effectively counter adversary information, it is necessary to understand the environment in which the adversary information exists. One way to establish the context of adversary information is to determine the interrelationship between information indigenous to the operational area and the culture and history of the people. This information is often available in the MISO studies and appendixes to the command's operation plan. Furthermore, it is also necessary to identify adversary information from other forms of information in the operating environment. Adversary information is often subtle and nuanced, and may be mixed in with misinformation and disinformation. To separate propaganda, it is necessary to identify adversary capabilities to develop and spread propaganda, as well as the receptiveness of the target audience to the adversary's lines of persuasion. This is typically a MISO task, conducted using the source-content-audience-media-effects analysis technique for individual pieces or instances of opponent propaganda and series analysis to determine the operational impact. Source-content-audience-media-effects analysis requires thorough analysis and resources normally found with the MIS elements at the JSOTF level. IO planners can facilitate this analysis by assisting intelligence and MIS personnel in the collection of suspected propaganda. A simple description of source-content-audience-media-effects is as follows:

- *Source.* Identify the originator or sponsor of the propaganda.
- *Content.* Identify the line(s) of persuasion used (the message and the source's desired effect).
- *Audience.* Identify the audiences targeted by the source and actually reached by the propaganda. This step is critical to countering adversary IO planning.
- *Media.* Identify the medium used and why that particular medium was selected by the source.
- *Effects.* Determine the impact of the opponent's propaganda on the target audience. Try to determine whether the propaganda has caused attitudinal or behavioral change.

2-78. A possible staff solution to the problem of countering adversary information activities is to form a working group of personnel from the IO, MISO, PA, and intelligence staffs who can fuse propaganda analysis and media analysis with the current intelligence estimate. In general, the working group seeks to determine how the adversary affects the content and flow of information in the operating environment, how propaganda impacts the various target audiences, and what audience needs are being targeted by the propaganda.

2-79. Countering adversary information activities does not commence upon discovery of adversary propaganda. Effective operations to counter adversary information activities proactively seek to mitigate propaganda's effects before their onset.

2-80. Countering adversary information activities is a long-term operation. To mitigate or nullify the effects of adversary propaganda, countermeasures must anticipate the adversary's response. Success of this effort rests with the ability to correctly direct the capabilities at affecting specific information to the target audience. An effective operation to counter adversary information efforts selects the appropriate capabilities and determines how these capabilities can be employed to match or overmatch the effects of opponent propaganda. Common techniques to counter adversary information include the following:

- *Forestalling.* Forestalling counters possible lines of persuasion prior to the release of propaganda.
- *Conditioning.* Conditioning preemptively shapes target audience vulnerabilities prior to exposure to propaganda.
- *Restrictive measures.* Restrictive measures deny the intended target audience access to the propaganda.
- *Direct refutation.* Direct refutation rebuts the propaganda point-for-point.

Information Operations Capabilities and Tactics

- *Indirect refutation.* Indirect refutation questions the validity of some aspect of the opponent's argument.
- *Diversion.* Diversion diverts attention by presenting more important or relevant themes to the target audience.
- *Imitative deception.* Imitative deception alters the propaganda to degrade its impact.
- *Silence.* Silence offers no response to the propaganda.
- *Minimization.* Minimization acknowledges selected elements of the propaganda while downplaying the importance of the content.

2-81. It is unlikely that any one set of countermeasures will apply a complete solution. The effects of opponent propaganda and friendly countermeasures will likely develop in a nonlinear fashion; hence, a constant process of analysis and application is necessary. It is unlikely that any one set of countermeasures will apply a complete solution. The effects of opponent propaganda and friendly countermeasures will likely develop in a nonlinear fashion; hence, a constant process of analysis and application is necessary. To do this, IO planners must monitor any effects produced by the countermeasures, changes to the operating and information environments, and adversary responses to the countermeasures. Then, if applicable, IO planners reengage the target audiences with new countermeasures. Although there is no doctrinal methodology for countering propaganda, the following steps can be used:

- *Analyze target audiences.* Understand the environment, the operational area, the inhabitants, the culture, and the adversary.
- *Analyze propaganda.* Establish a collection plan to identify and collect adversary propaganda. Use the source-content-audience-media-effects process to analyze.
- *Analyze media affecting the environment.* Identify media in AO and then determine its bias and use by adversary for propaganda purposes.
- *Apply countering adversary information measures.* Compare the propaganda analysis to the various capabilities and countering adversary information techniques and then apply appropriate countermeasures.
- *Monitor.* Evaluate the effects of the countering adversary information measures.

REWARDS PROGRAMS

2-82. The Department of Defense Rewards Program pays rewards to persons for providing USG personnel with information or nonlethal assistance that is beneficial to—

- An operation or activity of the Armed Forces or of allied forces participating in a combined operation with allied forces conducted outside of the United States against international terrorism.
- Force protection of the Armed Forces or allied forces participating in a combined operation with U.S. Armed Forces.

2-83. There are two types of Department of Defense rewards:

- Preapproved rewards allow a geographic combatant commander to nominate individuals or items to be placed on the Secretary of Defense preapproved rewards list for rewards in amounts that are in excess of the authority delegated to combatant commanders.
- Regular rewards paid to individuals providing the information after the target has been prosecuted and a monetary value established.

2-84. The Department of Defense Rewards Program can be used for information leading to the killing or capture of high-value individuals, the recovery of weapons caches, or information of impending attack on U.S. forces. Rewards can be paid in monetary funds or barter items. It cannot be used for weapons buyback programs, running an intelligence program, paying intelligence-source salaries, deceased persons (for example, an assassination program), or paying for illegal drugs (for example, a poppy, heroin, or cocaine buying program).

2-85. Rewards programs can be a potent asset to IO used to shape the information environment and provide a conduit to pass messages to the populace and the adversary. Examples of how rewards can be used for the purposes of IO include the following:

- *Key-leader engagement.* Rewards can be used as a means to establish working relationships and build influence with key leaders who have influence in their communities. Rewards can be used to bolster a key leader's position in their community.
- *Influence local population.* Rewards can be used to convince the local populace they can help control the local security situation. They can receive cash rewards for turning in weapon caches and insurgents that cause insecurity in their communities.
- *Message insurgent leaders and fighters.* Rewards can be used to send messages that affect adversary perceptions and decisionmaking. Placing a preapproved reward on a mid-level insurgent leader and then saturating his operating area with wanted posters and handbills may curtail his ability to move and conduct operations.
- *Rewards as part of deception.* Deceptive information about rewards can be used to conceal friendly TTP. Messages can be disseminated that insurgents are being captured because they are being turned in for rewards and not by friendly collection assets. The success of the reward program can be highlighted through MISO products, key-leader engagements, and the "rumor-mill."
- *Rewards as a divisive tool.* As rewards are paid to individuals concurrent with the kill or capture of high-value individuals, friction can be created within an enemy network as members consider who may be leaking information, intentionally or unintentionally, that places the network at risk.

CIVIL-MILITARY OPERATIONS

2-86. CMO establish, maintain, influence, or exploit relations between military forces, governmental and nongovernmental civil organizations, and the local populace. CMO contribute to shaping the operational area by focusing on civil aspects of the mission, their impact on military operations, and the impact of military operations on the civilian populace. A supportive civilian population can provide resources and information that facilitate friendly operations. As is the case with PA, CMO rely heavily on credibility with local leaders and the populace. CA forces are the designated forces and units organized, trained, and equipped to support the commander in planning and conducting CMO.

2-87. CA forces are structured to support JSOTF operations at the strategic, operational, and tactical levels while maintaining regional focus. The concept of CA support to the JSOTF is that a CA battalion (minus) with two CA companies supports the JSOTF. The CA battalion CA planning team is collocated with the JSOTF HQ to assist in CMO planning within the joint special operations area. A CA company HQ will be collocated with each SOTF and is capable of providing a civil-military operations center (CMOC) outside of each SOTF. A CA team is designated to support each SF advance operational base, as directed. The remaining CA teams are designated as a surge capability for the SOTF commander.

2-88. The CMOC is a standing capability formed by all CA units. The CMOC serves as the primary coordination interface for the U.S. armed forces and indigenous populations and institutions, humanitarian organizations, intergovernmental organizations, nongovernmental organizations, multinational military forces, and other civilian agencies of the USG. The CMOC facilitates continuous coordination among the key participants with regard to CMO and CAO from local levels to international levels within a given operational area, and develops, manages, and analyzes the civil inputs to the common operational picture. The CMOC center is the operations and support element of the CA unit as well as a mechanism for the coordination of CMO.

2-89. CA teams are typically four-Soldier elements consisting of a team leader, team sergeant, engineer, and medic. Besides planning, coordinating, and supporting civil reconstruction projects, CA teams can conduct medical civilian action programs, veterinary civilian action programs, and humanitarian assistance missions to provide quick-impact contributions to local populace quality of life. A CA teams' performance in a given area can affect the public perception in the local area. MISO can support these with radio broadcasts and other means to advertise the events and later exploit their success. PA can also exploit the success of these missions through press releases. It may be useful to attach COMCAM to a CA team to document civil contributions or to provide photographs for intelligence analysis. If COMCAM assets are not available, any available Soldier with knowledge of the command's intent for the photographs may take pictures.

2-90. If friendly-force military operations create collateral damage and casualties among the populace, CA teams may conduct consequence management to mitigate the negative impact of such operations on the populace through payments or other types of reimbursement. The unit's CA officer may be a conduit to a provincial reconstruction team or other similar entities that have developed relationships with local leaders.

PUBLIC AFFAIRS

2-91. PA units provide timely and accurate information so that both U.S. and international audiences may assess and understand the facts concerning military operations. PA units have the following responsibilities:

- *Internal.* PA provides command information to inform the force and counter effects of adversary propaganda and misinformation.
- *External.* PA provides information regarding military operations to external agencies, governments, media, and populaces.

2-92. Although PA personnel strive to be separate from IO, there is no denying the impact of timely, relevant PA press releases on a given audience. This impact can be multiplied substantially and used to support command objectives when PA personnel are included in IO planning and able to prepare consequence-management activities by preparing press releases beforehand.

2-93. Adversary forces may review PA release information to cue intelligence and provide battle damage assessments. Additionally, PA may be an information conduit to adversary decisionmakers. Extreme care should be taken when employing PA, as successful and effective public relations depends on credibility, and credibility relies on truthful reporting. PA can support IO by—

- Getting ahead of enemy propaganda with the truth.
- Countering adversary misinformation and disinformation by publishing accurate information.
- Ensuring media awareness of the implications of premature release of certain information.
- Playing a key role in establishing ground rules for embedded reporters.

2-94. MISO and PA are separate capabilities that support the commander's objectives. Coordination must be conducted between both. For IO planners, it is important to understand that rural populations generally do not have access to PA release material. Further, PA is wholly dependent on local, regional, and international media to carry their messages. These rural and isolated groups are best informed through MIS assets, which do have organic production and dissemination capabilities. IO planners can facilitate this by ensuring PA releases and articles are sent to MISO planners for dissemination to the local populace. PA planners should review and deconflict messaging with MISO planners to ensure that tactical, operational, and strategic messages are mutually supporting or, at the very least, not contradictory.

> *Note.* The best way to influence the populace through themes and messages is to use the most influential people in the area—the key communicators. Local leaders, FID partner forces, and religious leaders are just some examples of the types of personnel who can spread the message amongst the populace. Whoever is viewed as the most influential, trustworthy source (key is credibility of the messenger) should be the individual serving as a conduit to the target audience.

DEFENSE SUPPORT TO PUBLIC DIPLOMACY

2-95. DSPD are those activities and measures taken by Department of Defense components to support and facilitate public diplomacy efforts of the USG. DSPD is a key military role in supporting the USG's strategic communication program. It includes peacetime military engagement activities conducted as part of the combatant commanders' theater security cooperation plans.

2-96. The focus of defense support to the public diplomacy is to understand, engage, influence, and inform critical foreign audiences through words and actions to foster understanding of U.S. policy and advance U.S. interests. The decisions and actions executed by the SFODA on the ground can have a profound effect on U.S. public diplomacy efforts.

2-97. Figure 2-11, pages 2-32 through 2-35, and Figure 2-12, pages 2-36 through 2-38, provide an overview of IO capabilities. Figure 2-13, pages 2-39 and 2-40, outlines the support roles of IO, CMO, and PA.

	OPSEC Supports By:
MILDEC	• Concealing competing observables. • Degrading general situation information to enhance effect of observables. • Limiting information and indicators that could compromise military deception operations.
MISO	• Concealing contradicting indicators while conveying selected information and indicators. • Ensuring products do not contain classified information.
Physical Destruction	• Concealing friendly delivery systems from enemy offensive IO until it is too late for the adversary to react. • Denying information to the enemy on the success of offensive IO.
EW	• Concealing EW units and systems to deny information on extent of EA and EW support capabilities.
Physical Security	• Concealing EEFI. • Reducing the activities requiring physical security. • Hiding tools of physical security, thus preventing adversary from gaining access.
IA	• Concealing physical and electronic information system locations.
CI	• Ensuring EEFI are concealed from enemy collection assets.
Computer Network Attack (CNA)	• Concealing CNA capabilities.
CND	• Denying enemy knowledge about CND capabilities.
	MILDEC Supports By:
OPSEC	• Influencing adversary not to collect against protected units/activities. • Causing adversary to underestimate friendly operations security capabilities.
MISO	• Providing information compatible with MISO theme.
Physical Destruction	• Influencing adversary to underestimate friendly physical-destruction capabilities. • Influencing adversary to defend C2 element/systems that friendly forces do not plan to destroy.
EW	• Influencing adversary to underestimate friendly EA and EW support capabilities.
Physical Security	• Masking troop activities requiring safeguards.
IA	• Overloading adversary intelligence and analysis capabilities. • Protecting and defending friendly information systems.
CI	• Giving the adversary a cover story so his intelligence system collects irrelevant information.
CNA	• Providing MILDEC targets and deception stories to enhance CNA.

Figure 2-11. Mutual support within information operations capabilities

	MILDEC Supports By (continued):
CND	• Causing the enemy to believe U.S. CND defense is greater than it actually is. • Causing the enemy to believe all CND tools are in place.
MISO Supports By:	
OPSEC	• Disseminating rules of engagement. • Assisting in the countering of propaganda and misinformation. • Minimizing resistance and interference by local population.
MILDEC	• Creating perceptions and attitudes that MILDEC can exploit. • Integrating MISO actions with MILDEC. • Reinforcing the deception story with information from other sources.
Physical Destruction	• Causing populace to leave targeted areas to reduce collateral damage.
EW	• Broadcasting MISO products into adversary civilian and military frequencies. • Developing messages for broadcast on other service EW assets.
Physical Security	• Targeting adversary audiences to reduce the need for physical security.
IA	• Enhancing the ability of IA in the minds of the enemy.
CI	• Providing messages in enemy decisionmaker's mind that can be revealed by CI to determine enemy true intentions.
CNA	• Convincing enemy to not do something by describing effects of a CNA if they take undesirable actions. • Providing MISO messages for dissemination by CNA means.
CND	• Providing information about nonmilitary threat to computers in the AO.
Physical Destruction Supports By:	
OPSEC	• Preventing or degrading adversary reconnaissance and surveillance.
MILDEC	• Conducting physical attacks as deception events.
MISO	• Degrading adversary's ability to see, report, and process information. • Isolating target audience from information.
EW	• Destroying adversary C2 targets.
Physical Security	• Reducing physical security needs by attacking adversary systems able to penetrate information systems.
IA	• Attacking adversary systems capable of influencing friendly information systems availability and integrity.
CI	• Destroying appropriately nominated adversary collection assets.
CNA	• Supplementing computer network attack by destroying or degrading hard targets.
CND	• Destroying or degrading enemy CNA facilities before they attack friendly computers.
EW Supports By:	
OPSEC	• Degrading adversary electromagnetic intelligence, surveillance, and reconnaissance operations against protected units and activities. • Creating barrier of white noise to mask unit maneuvers.
MILDEC	• Using EA and EW support as deception measures. • Degrading adversary capabilities to see, report, and process competing observables causing the enemy to misinterpret information received by electronic means.

Figure 2-11. Mutual support within information operations capabilities (continued)

Chapter 2

	EW Supports By (continued):
MISO	• Degrading adversary's ability to see, report, and process information. • Isolating target audience from information and herding that target audience onto MISO broadcast frequencies.
Physical Destruction	• Providing target acquisition through EW support.
Physical Security	• Using electronic protection to safeguard communications used in protecting facilities.
IA	• Using electronic protection to protect equipment.
CI	• None.
CNA	• Supplementing CNA with EA.
CND	• Using electronic protection to protect personnel, facilities, and equipment.
	Physical Security Supports By:
OPSEC	• Protecting operation plans and operation orders.
MILDEC	• Restricting access by level of security and number of personnel.
MISO	• Protecting inventory of sensitive products to prevent premature dissemination of messages.
Physical Destruction	• Safeguarding availability of information systems to use in physical destruction.
EW	• Safeguarding equipment used in electronic warfare.
IA	• Safeguarding information systems by implementing security procedures.
CI	• Safeguarding personnel, and preventing unauthorized access to equipment, installation, materiel, and documents.
CNA	• Safeguarding information systems from sabotage, espionage, damage, or theft.
CND	• Determining applicable risk and threat levels.
	IA Supports By:
OPSEC	• Ensuring information system confidentiality.
MILDEC	• Providing information system assets for conducting MILDEC operations.
MISO	• Ensuring availability of information systems for MISO.
Physical Destruction	• Ensuring information systems are available for physical destruction tasks.
EW	• Ensuring EW assets are available.
Physical Security	• Providing for information system authentication.
CI	• Ensuring information systems are available to conduct CI.
CNA	• Ensuring links with higher HQ to pass CNA.
CND	• Taking actions to ensure availability, integrity, authentication, confidentiality and nonrepudiation of computer.
	CI Supports By:
OPSEC	• Countering foreign human-intelligence operations.
MILDEC	• Countering foreign human-intelligence operations. • Identifying threat intelligence, surveillance, and reconnaissance capabilities.
MISO	• None

Figure 2-11. Mutual support within information operations capabilities (continued)

Information Operations Capabilities and Tactics

	CI Supports By (continued):
Physical Destruction	• None.
EW	• Providing electronic countermeasures. • Conducting countersignal operations to allow broadcast of MISO messages.
Physical Security	• Countering foreign human-intelligence operations.
IA	• At certain echelons, helping ensure information integrity.
CNA	• Confirming results of CNA.
CND	• Detecting, identifying, assessing, countering, and neutralizing enemy intelligence collection.
	CNA Supports By:
OPSEC	• Attacking enemy computers before they can detect U.S. EEFI.
MILDEC	• Providing the deception story through computers.
MISO	• None.
Physical Destruction	• Attacking selected targets by nonlethal means, which allows lethal attacks on other targets.
EW	• Using with EA.
Physical Security	• Conducting risk assessment to determine consequence of second- and third-order computer network attack effects.
IA	• Attacking enemy computers before the enemy attacks friendly computers.
CI	• Exploiting enemy intelligence collection.
CND	• Attacking the enemy's ability to attack friendly computers.
	CND Supports By:
OPSEC	• Detecting enemy attempts to acquire information.
MILDEC	• Protecting the MILDEC plan resident inside computers.
MISO	• Preventing the compromise of MISO message before release.
Physical Destruction	• Protecting fire support C2 systems.
EW	• Using in conjunction with electronic protection.
Physical Security	• Erecting firewalls to protect intrusion into networks.
IA	• Supporting information assurance of information passed via computer networks.
CI	• Detecting, identifying, and assessing enemy collection efforts against computers.
CNA	• Protecting CNA weapons from enemy detection.

Figure 2-11. Mutual support within information operations capabilities (continued)

Chapter 2

	OPSEC Can Conflict By:
MILDEC	• Limiting information that can be revealed to enhance deception story credibility.
MISO	• Limiting information that can be revealed to develop MISO messages.
Physical Destruction	• Limiting information that can be revealed to enemy to develop targets.
EW	• Electronic protection and operations security may have different goals.
Physical Security	• Should be no conflict.
IA	• Should be no conflict.
CI:	• Should be no conflict.
CNA	• Should be no conflict.
CND	• Should be no conflict.
	MILDEC Can Conflict By:
OPSEC	• Revealing information OPSEC normally seeks to conceal.
MISO	• Limiting MISO theme selection. • Limiting information that can be revealed to develop military information themes. • Undermining the credibility of overt messages and other MISO efforts.
Physical Destruction	• Limiting targeting to allow survival and conduct of critical adversary C2 functions.
EW	• Limiting EA targeting of adversary information systems to allow survival and conduct of critical adversary C2 functions.
Physical Security	• Negating the deception story by physical security preventing transmission of a realistic deception story.
IA	• Presenting data the enemy will believe versus assuring data is not revealing to enemy.
CI	• Giving the adversary a cover story that inadvertently supports his collection plan.
CNA	• Should be no conflict.
CND	• Should be no conflict.
	MISO Can Conflict By:
OPSEC	• Revealing information OPSEC normally seeks to conceal.
MILDEC	• Limiting deception story selection if deception story contains untruths.
Physical Destruction	• Limiting targeting of adversary C2 infrastructure to allow conveying of MISO messages.
EW	• Limiting EA against adversary communications frequencies to allow MISO messages to be conveyed.
Physical Security	• Should be no conflict.
IA	• Should be no conflict.
CI	• Should be no conflict.
CNA	• Should be no conflict.
CND	• Should be no conflict.

Figure 2-12. Potential conflicts within information operations capabilities

Physical Destruction Can Conflict By:	
OPSEC	• Causing firing systems to reveal their locations.
MILDEC	• Limiting selection of deception means by denying or degrading elements of adversary C2I command infrastructure necessary to process deception story.
MISO	• Limiting means available to convey MISO messages by denying or degrading adversary C2 systems and civilian communications infrastructure.
EW	• Limiting opportunities for communications intrusion by denying or degrading elements of adversary information systems.
Physical Security	• Limiting access to targeting data (consider need to know).
IA	• Attacking incorrect adversary systems capable of influencing friendly information system availability and integrity.
CI	• Destroying insufficient number of adversary collection assets.
CNA	• Should be no conflict.
CND	• Should be no conflict.
EW Can Conflict By:	
OPSEC	• Revealing EW assets prematurely.
MILDEC	• Limiting selection of deception measures by denying or degrading use of adversary C2 systems.
MISO	• Reducing frequencies available to convey MISO messages. • Jamming military and commercial frequencies used by MISO for electronic dissemination.
Physical Destruction	• Limiting targeting of adversary C2 systems.
Physical Security	• Revealing what physical security is trying to protect (EA). • Electronic protection should not conflict.
IA	• Should be no conflict.
CI	• Should be no conflict.
CNA	• Should be no conflict.
CND	• Should be no conflict.
IA Can Conflict By:	
OPSEC	• Should be no conflict.
MILDEC	• Reinforcing the deception story.
MISO	• Should be no conflict.
Physical Destruction	• Should be no conflict.
EW	• Deconflicting electronic protection and information assurance.
Physical Security	• Should be no conflict.
CI	• Having insufficient information systems available to conduct CI.
CNA	• Having no available links with higher HQ to pass CNA requests.
CND	• Should be no conflict.

Figure 2-12. Potential conflicts within information operations capabilities (continued)

Chapter 2

	CI Can Conflict By:
OPSEC	• Should be no conflict.
MILDEC	• Should be no conflict.
MISO	• Should be no conflict.
Physical Destruction	• Killing sources.
EW	• Needing EW support for other activities.
Physical Security	• Should be no conflict.
IA	• Negating information integrity with ineffective CI.
CNA	• Should be no conflict.
CND	• Revealing CI on how networks are protected.
	CNA Can Conflict By:
OPSEC	• Attacking selected enemy targets may provide information on friendly activities.
MILDEC	• Resulting in attacks on wrong target if coordination is not made with MILDEC.
MISO	• Preventing the enemy from receiving MISO messages.
Physical Destruction	• Attacking same target with nonlethal and lethal weapons wastes both time and ammunition.
EW	• Needing to deconflict which systems attack which targets.
Physical Security	• Revealing computer network attack sources that should be protected.
IA	• Should be no conflict.
CI	• Attacking enemy computers before exploiting hostile intelligence collection efforts.
CND	• Should be no conflict.
	CND Can Conflict By:
OPSEC	• Should be no conflict.
MILDEC	• Reinforcing the deception story.
MISO	• Should be no conflict.
Physical Destruction	• Should be no conflict.
EW	• Should be no conflict.
Physical Security	• Should be no conflict.
IA	• Should be no conflict.
CI	• Should be no conflict.
CNA	• Should be no conflict.

Figure 2-12. Potential conflicts within information operations capabilities (continued)

	IO Supported By:
CMO	- Influencing informing populace of civil-military activities and support. - Neutralizing misinformation and hostile propaganda directed against civil authorities. - Controlling electromagnetic spectrum for legitimate purposes.
PA	- Countering adversary information and protecting from misinformation/rumor. - Developing EEFI to preclude inadvertent public disclosure. - Synchronizing MISO and OPSEC with PA strategy.
DSPD	- Ensuring accuracy of information. - Maintaining relevance of information. - Timeliness of information. - Usability of information. - Completeness of information. - Security of information.
COMCAM	- Coordinating guidance to COMCAM teams with commander's information/objectives. - Assisting in expeditious transmission of critical COMCAM images.
	CMO Supported By:
IO	- Providing information to support friendly knowledge of information environment. - Synchronizing communications media and assets and messages with other information capabilities. - Coordinating C2 target sets with targeting cell. - Establishing and maintaining liaison or dialogue with indigenous personnel and nongovernmental organization. - Supporting MISO with feedback on MISO themes. - Providing news and information to the local people.
PA	- Providing information on CMOC activities to support PA strategy. - Synchronizing communications, media, and message. - Identifying, coordinating, and integrating media, public information, and HN support.
DSPD	- Providing information to inform interagency elements on local information environment. - Synchronizing communications media and messages with other IO capabilities. - Establishing and maintaining liaison or dialogue with indigenous personnel and nongovernmental organizations. - Supporting DSPD with feedback on strategic communications themes.
COMCAM	- Using COMCAM capabilities to record priority civic action projects. - Synchronizing imagery assignments with COMCAM team leader.
	PA Supported By:
IO	- Coordinating with IO planners to ensure a consistent message and maintain OPSEC. - Supporting counter adversary information. - Providing assessment of effects of media coverage to OPSEC planners. - Providing assessment of essential nonmedia coverage of deceptions story.

Figure 2-13. Support roles of information operations, civil-military operations, public affairs, defense support to public diplomacy, and combat camera

	PA Supported By (continued):
CMO	• Providing accurate, timely, and balanced information for the public. • Coordinating with civil affairs specialist to verify facts and validity of information.
DSPD	• Coordinating with interagency planners to ensure a consistent message. • Proving assessment of media coverage.
COMCAM	• Managing release of key images through PA channels. • Coordinating for COCAM coverage and access to key events and operation.
	DSPD Supported By:
IO	• Providing a link to interagency for coordination and guidance on strategic communications themes and activities.
CMO	• Providing a link to interagency for coordination and guidance on strategic communications themes and activities.
PA	• Providing a link to interagency for coordination and guidance on strategic communications themes and activities.
COMCAM	• Providing a link to interagency for coordination and guidance on strategic communications themes and activities.
	COMCAM Supported By:
IO	• Providing responsive imagery coverage of events in the operational area.
CMO	• Providing responsive imagery coverage of events in the operational area.
PA	• Providing responsive imagery coverage of events in the operational area.
DSPD	• Providing responsive imagery coverage of events in the operational area.

Figure 2-13. Support roles of information operations, civil-military operations, public affairs, defense support to public diplomacy, and combat camera (continued)

Chapter 3

Planning Information Operations

IO are planned as part of the planning process, whether it is the military decisionmaking process, the joint planning process, or some abbreviated planning method. However, in comparison with planning other operations, there are two noticeable differences:
- A longer lead time is required for planning IO. Many IO capabilities have time requirements for preparation (notably MILDEC, MISO, and CNO).
- The threat of hostile information from outside the operational area is great. The ease of information flow through information networks and the media means that operating boundaries are porous to outside influences.

The focus of mission planning for IO is to gain information superiority. Information superiority is an operational advantage derived from the ability to collect, process, and disseminate an uninterrupted flow of information while exploiting or denying an adversary's ability to do the same. Because absolute information superiority is rarely possible to gain or maintain, IO should seek information superiority at or before the operation's decisive point. If the operation is phased, it may be necessary to achieve a form of information superiority in each phase. If the operation is not phased, planners may determine that operational advantages are needed before, during, and after the operation.

Successful IO Plan

The primary role of the IO planner is to coordinate, synchronize, and deconflict while ensuring the appropriate capabilities are employed based on the desired effect. An example of this is an information operation that took place in the Basra Province of Iraq in 2009.

The mission was to conduct an information operation with and through the HN forces to bolster popular support for a partner unit and decrease the influence of extremist groups in the province to set conditions for a safe and secure environment.

The concept was broken down into three phases and focused on four information capabilities: PSYOP, PA, COMCAM, and CMO. The first phase was education and training; the focus of this phase was to provide the partner unit with training on how to interact properly with the populace and the media. The PSYOP teams took the lead during this phase by providing information team training based on various approved courses. This was also an assessment phase for the PAO and the CA officer on the partner unit's media relations and CMO capabilities. In addition to conducting assessments, the PAO, in cooperation with COMCAM, published numerous press releases highlighting the successes of the partner unit's direct action missions. The second phase was planned engagements; the focus of this phase was to assist the partnered unit with providing the locals with much-needed supplies and materials. The CA planner took the lead during this phase by working with the SFODA to determine the best areas to deliver goods and in which areas the SFODA wanted to increase their influence. *(continued)*

> The PAO and COMCAM supported this phase by highlighting the engagements via press releases and photos. They also utilized a broadcaster to gather video footage of the successful engagements. Based on the relationships developed during phase two, the SFODA transitioned to phase three—planned operations. The SFODA used the information gathered during the planned engagements to assist with targeting efforts, and focused on decreasing the influence of the insurgent groups in the region.
>
> By incorporating multiple capabilities of IO into a focused, coordinated effort, the plan achieved the desired effects to bolster popular support for a partner unit and decrease the influence of extremist groups in the province.

THE STAFF ESTIMATE FOR INFORMATION OPERATIONS

3-1. The staff estimate is an assessment of the situation and an analysis of the COAs a commander is considering. It includes an evaluation of how factors in a staff section's functional area influence each COA or assigned mission, and includes conclusions and recommendations. Staff estimates are developed as part of the planning process. These estimates normally are text documents; however, they may be formatted as maps, graphics, or charts. Whatever form they take, the estimate should be as comprehensive as possible without becoming overly time-consuming.

3-2. The staff estimate for IO is an estimate focused on the information environment and the use of information by adversary and friendly forces. It assesses the situation in the information environment and analyzes the best way to achieve information superiority for the assigned mission.

3-3. Staff sections, particularly at the tactical level, rarely have the time to complete all five paragraphs of a formal doctrinal estimate. In that case, the estimate should concentrate on situation assessment rather than COA development, and only paragraphs 1 and 2 need be produced and updated as operations progress.

3-4. Figure 3-1, page 3-3, provides a format for an IO staff estimate. Figure 3-2, page 3-4, provides an example of a graphic IO estimate.

PLANNING CONSIDERATIONS DURING MISSION ANALYSIS

3-5. The purpose of mission analysis for IO is to assist planners in seeing the information environment, the adversary, and friendly forces in the context of the assigned mission. At the end of mission analysis, IO personnel should have—

- The intelligence preparation of the operational environment (IPOE) products, such as a combined information overlay (CIO) and a template of adversary operations in the information environment.
- The essential tasks for IO.
- The capabilities in the information environment.
- The constraints for IO.
- The critical information requirements for IO.
- The EEFI.

ESSENTIAL TASKS FOR INFORMATION OPERATIONS

3-6. Units rarely conduct an information operation autonomously. There will always be higher HQ and tasks. Although some tasks may have been specifically assigned by the higher HQ, others may be implied (meaning they are necessary to accomplish specified tasks or the overall mission). Implied tasks should require resources and not be administrative in nature. From the specified and implied tasks, planners should identify tasks that the command must successfully accomplish to affect adversary and friendly use of information. These are the unit's essential tasks for IO.

1. **MISSION.** The unit mission.
2. **SITUATION AND CONSIDERATIONS.**
   a. **Characteristics of the Information Environment.** Summarize significant characteristics of the information environment and the impact on military operations.
      (1) Subinformation environments. How terrain and weather, populace, civilian information infrastructure, civilian population, third-party organizations, and other physical and cognitive features of the information environment create subinformation environments.
      (2) Information nodes. Identify what places, persons, or infrastructure in each subinformation environment shape information contact and flow by creating or transmitting information.
   b. **Adversary Forces.** Adversary capabilities, vulnerabilities, and activities in the information environment.
   c. **Friendly Forces.**
      (1) Friendly COA. IO concept of support for each COA.
      (2) Current status of resources. The availability of organic IO capabilities and assets (as translated into capabilities to operate in the information environment).
      (3) Current status of other resources. The availability of supporting IO capabilities and assets from higher HQ, other commands, agencies, and organizations.
      (4) Friendly-force vulnerabilities in the information environment.
      (5) Comparison of requirements versus capabilities and recommenced solutions.
      (6) Key considerations (evaluation criteria) for COA supportability.
   d. Assumptions. Assumptions for IO developed during mission analysis.
3. **COURSES OF ACTION.**
   a. List the COAs that were war-gamed.
   b. List evaluation criteria identified during COA analysis.
4. **COA ANALYSIS.** Analyze each COA using the evaluation criteria. Estimate the likelihood of accomplishing the IO objectives given the available time and capabilities. Determine the potential for unintended consequences of IO tasks and the possible impacts on friendly and adversary forces' COAs.
5. **COMPARISON.** Compare COAs using evaluation criteria. Rank-order COAs for each criterion. If possible use a decision matrix to support.
6. **CONCLUSIONS AND RECOMMENDATION.** Recommend COA based on the comparison (most supportable by IO). Identify IO issues, deficiencies, risks, and recommendations to reduce their impacts.

Figure 3-1. Information operations staff estimate

3-7. Typically, essential tasks for IO number between three and five. More than five essential tasks present the risk of overtaking subordinate elements or having an information operation that is too complex to execute.

3-8. One useful technique for validating an essential task is to ask the following question: "If the unit accomplishes all other tasks marginally and does this one well, will it accomplish the mission?" If the answer is "no," then the task is not essential. If more than five essential tasks are identified, planners should question the validity of each essential task or the nature of the requirements levied on the unit by higher HQ.

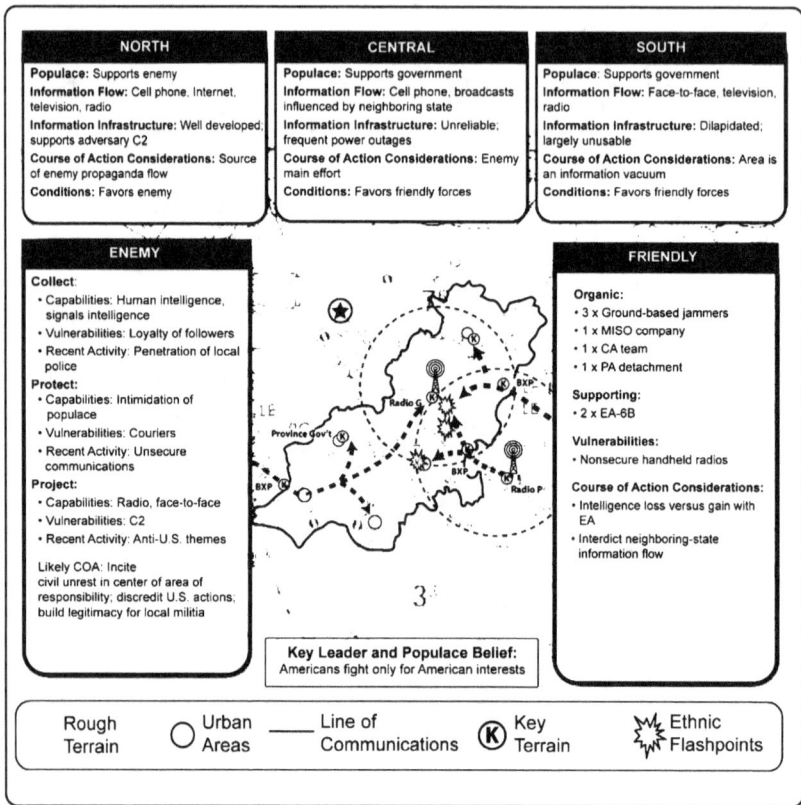

Figure 3-2. Example graphic information operations estimate

CAPABILITIES IN THE INFORMATION ENVIRONMENT

3-9. IO planners should determine if the command has the assets to perform assigned tasks. This is done by identifying any or all organic and supporting IO-capable assets. Organic assets are resident in assigned or attached forces. Supporting assets are available to the command from a higher HQ or government agency. Available assets are then compared with the IO mission requirements (specified and implied tasks) to identify capability shortfalls and any additional assets that are required. It is important to keep in mind that IO planners do not possess any of the individual IO capabilities; these assets all reside within the supporting elements. To ensure use of these assets, IO personnel must start coordination early.

Planning Information Operations

3-10. IO planners face a challenge in expressing IO capabilities to the commander and staff. A simple list of IO capabilities (that is, three ground-based jammers, nine MIS teams, two COMCAM teams, and so on) does not help the commander visualize the command's capabilities in the information environment. In developing its staff estimate, IO planners should consider the following basic questions:
- What can the command do using organic assets?
- What can supporting assets from the higher HQ do?
- What can not be done?

3-11. One possible solution to this problem is to organize IO capabilities by elements, asset, and means in terms of their contributions to friendly-force operations in the information environment. Figure 3-3 shows a sample asset list.

Element	Assets	Means	Supported Essential Tasks for IO	Effect	Targets
MIS detachment	• 4 x MIS teams	• Loudspeaker operations • Handbills and posters • Radio and TV broadcasts • Face-to-face • Key-leader engagement	• Degrade adversary morale • Influence local populace to not interfere with friendly operations	• Degrade • Influence • Inform	• Insurgent cell • Local populace • Key leaders
SF detachment	• SF teams	• Direct action • Face-to-face	• Disrupt adversary C2 • Build legitimacy of HN security forces	• Disrupt • Destroy • Degrade • Influence	• Insurgent leaders • Insurgent cells • Local populace
EA	• 2 x EA-6B	• Jammer	• Disrupt adversary C2 • Degrade adversary morale	• Disrupt • Degrade • Isolate • Influence	• Insurgent leaders • Insurgent cells

Figure 3-3. Sample information operations asset list

CONSTRAINTS ON INFORMATION OPERATIONS

3-12. Constraints are restrictions placed on the command by a higher HQ that either require the use of resources to execute a specific task or prohibit the commander from taking specific actions. In either case, constraints reduce the commander's freedom of action.

3-13. Like most other operations, IO are constrained by rules of engagement, U.S. national policy, international politics, and other legal, moral, cultural, or operational factors. Additionally, IO planners should consider that IO capabilities have constraints of their own, particularly MILDEC, MISO, CNO, and EW. Common constraints include approval authority for deception operations, MISO themes to avoid, allied forces' national policies and capabilities, restricted targets and frequencies, and PAO guidance.

3-14. To enhance understanding, IO constraints are organized in terms of information content and flow, as shown in Figure 3-4, page 3-6.

Chapter 3

Information Content	• Themes should avoid favoring any ethnic group. • Themes should stress highlighted cooperation. • During current operations, approval authority is delegated down to the JSOTF and brigade combat team commanders. • Joint task force commander approves deception.
Information Flow	• No cross-boundary EA. • All EA must be coordinated with the JSOTF. • Non-U.S. allies may not disseminate U.S. MISO products. • PA posture is passive. • Mosques are on the restricted-target list. • COMCAM priorities.

Figure 3-4. Example of information content and flow organization

CRITICAL INFORMATION REQUIREMENTS FOR INFORMATION OPERATIONS

3-15. Commander's critical information requirements (CCIRs) identify information needed by the commander to visualize the operational area and make critical decisions. CCIRs also filter information to the commander by defining what is important to mission accomplishment. If the information operation is important to the mission, then there should be IO input to the CCIR.

3-16. The staff nominates information requirements to become CCIRs based upon the commander's guidance, higher HQ CCIRs, the essential-task list, and the J-2/G-2/S-2 IPOE (situation template). There are two types of CCIRs—
- *Priority intelligence requirements (PIRs)*. PIRs are information the commander must know about the adversary. For IO, PIRs should focus on conditions in the information environment and adversary actions that affect the information environment. PIRs that may be required for IO include the following:
    - What media outlets are producing or disseminating hostile propaganda?
    - What propaganda themes are being disseminated to the populace by adversary forces?
- *Friendly-force information requirements (FFIRs)*. FFIRs are items of information the commander must know about the friendly force. For IO, FFIRs provide information on critical aspects of the command's information system, IO-capable assets, and execution of the information operation. FFIRs that may be required for IO include the following:
    - Death or serious injury of noncombatants by friendly forces.
    - Media coverage of alleged friendly-force misconduct.

ESSENTIAL ELEMENTS OF FRIENDLY INFORMATION

3-17. EEFI are the critical aspects of a friendly operation that—if known by the adversary—would subsequently compromise, lead to failure, or limit success of the operation, and therefore must be protected from detection. In other words, EEFI is a list of information that must be protected from the adversary's intelligence system to prevent the adversary from making timely decisions and allowing friendly forces to retain the initiative. Typically, EEFI include the command intentions, subordinate element status, or the location of critical assets (such as command posts and signal nodes). EEFI should be refined throughout the planning process, as some information may not be identified until COA development. Once EEFI are developed, specific measures (in the form of tasks to subordinate units) are developed to protect the information (OPSEC process). Two examples of EEFI are—
- Friendly forces' means of intelligence collection.
- Tribal leaders assisting friendly forces.

Note. Chapter 4 provides additional information on how to develop EEFI.

Planning Information Operations

MISSION ANALYSIS WORK SHEET

3-18. A mission analysis work sheet (Figure 3-5) guides planners through the critical parts of mission analysis. The format to conduct a mission analysis brief to the commander is identical to the mission analysis brief format.

---

1. **Facts.** Statements of known data concerning the situation, including adversary and friendly disposition, available troops, unit strengths, and material readiness that will directly affect the mission.
2. **Assumptions.** Suppositions on the current or future situation assumed to be true in the absence of facts and which will typically describe future eventualities on which success of the operation depends.
3. **Tasks.**
   a. **Specified.** Tasks specifically assigned to the command by higher HQ (extracted from paragraphs 1, 2, and 3 of the higher HQ base order, annexes, and overlays).
   b. **Implied.** Tasks that must be performed to accomplish specified tasks of the overall mission (developed from an analysis of specified tasks).
   c. **Essential.** Tasks that must be executed to accomplish the mission (derived from analyzing specified and implied tasks lists: essential tasks are included in the mission statement).
4. **Constraints.** Restrictions placed on the command by higher HQ that dictate an action or inaction, thus restricting the command's freedom of action (extracted from higher HQ guidance, concept of operations, coordinating instructions, and annexes—especially IO, rules of engagement, themes to avoid, CMO, and PA).
5. **Available Assets.** Organic and supporting troops and equipment available for the operations (derived from higher HQ order, current tasks organization, and unit status reports).
6. **Risk Assessment.** Hazards that may be encountered during the mission because of the presence of the adversary or hazardous condition in the AO (developed from staff experience and SOPs).
7. **CCIRs.** Information the commander needs to make critical decisions, especially to determine or validate COAs.
   a. **PIR.** Information the commander must know about the adversary (derived from known gaps in information required to accomplish the operations).
   b. **FFIRs.** Information the commander must know about the friendly force (developed form knowledge of the friendly force and mission).
8. **EEFI.** Critical aspects of the friendly operation that—if known by the adversary—will compromise, lead to failure, or limit success of the operation and, therefore, must be protected from detection (derived from higher HQ order and developed by using the OPSEC process).

Figure 3-5. Mission analysis work sheet

COURSE OF ACTION DEVELOPMENT

3-19. A COA is a possible plan to accomplish the assigned mission. The IO planner's goal is to develop a concept of support that will generate effects that create information superiority over the adversary at the proper time and place. An IO concept of support should be—
- *Suitable.* The concept must create information superiority over the adversary.
- *Feasible.* The COA must be practical in terms of time, space, and resources. Considerations include time available to shape the information environment and availability of IO capabilities.
- *Acceptable.* The command's information operation must consider the cost of resources, as well as the operational and accidental risks associated with the proposed concept.

- *Distinguishable.* Each COA should be supported with a unique information operation, although the differences may be subtle. These differences include the use of different IO capabilities, changes to the allocation of the capabilities, or changes in the time or sequence of IO tasks.
- *Complete.* The COA should provide information superiority and address friendly-force actions in the case of undesirable results.

3-20. Successful IO give subordinates maximum latitude for initiative and postures the unit for follow-on missions. Likewise, with a little foresight, IO planners can use one information operation to jump-start another. Occasionally, a tactical-level information operation may just be the perfect catalyst for an operational-level information operation (and so on).

STRENGTHS AND WEAKNESSES: INFORMATION ADVANTAGE

3-21. The first step in developing an IO concept of support is to determine whether friendly or adversary forces have the information advantage. A nondoctrinal term, *information advantage* means being in a superior position (able to operate better) in the information environment relative to one's opponent. Information advantage is relative, meaning that although two opposing forces are operating in the same information environment, how each force operates in the information environment is different.

3-22. To determine relative information advantages, leaders compare friendly and adversary forces' strengths (capabilities) and weaknesses (vulnerabilities) in the information environment. This analysis is an asymmetric evaluation, meaning that there will unlikely be a direct correlation between the assets used by either friendly or adversary forces, and how those forces employ the assets. Therefore, IO planners should attempt to relate similar capabilities and attributes in terms of how information is—

- *Collected.* Information collection describes how friendly and adversary forces' means and capabilities are used to collect information about the opponent. Leaders must consider capabilities in terms of HUMINT, SIGINT, imagery intelligence, measurement and signature intelligence (MASINT), and OSINT.
- *Protected.* Information protection describes friendly and adversary forces' means and capabilities to protect critical information and maintain means of communication.
- *Projected.* Information projection describes friendly and adversary forces' means and capabilities to put information into the operational area's information environment. Leaders must consider the type and number of information systems possessed by each side (for example, face-to-face, radio, or TV).

3-23. The adversary forces' capabilities and weaknesses can be derived from IPOE, whereas friendly-force capabilities come from mission analysis, and vulnerabilities from a center-of-gravity (COG) analysis of friendly forces. The results are compared to determine which side is at an advantage or disadvantage in each function. If no apparent advantage or disadvantage exists, then that aspect of operations in the information environment is neutral for both sides. The end result is a subjective determination of whether friendly or adversary forces have the overall advantage in the information environment, and in what way one side has the advantage. Once the analysis is concluded, IO planners should have insight into the following:

- The friendly forces' information capabilities needed for the operation.
- The friendly and adversary forces' vulnerabilities in the information environment.
- The type of operations in the information environment that may be possible from both the friendly and adversary forces' perspectives.
- The additional IO capabilities and resources required to execute the operation.
- The allocation of existing IO capabilities and resources.

The information advantage work sheet (Figure 3-6, page 3-9) is a tool for estimating information advantage.

Planning Information Operations

Capabilities in the Information Environment	Strength/Weakness		Relative Advantage	
	Enemy	Friendly	Enemy	Friendly
Information Collect	• OSINT • Couriers	• Overhead and night capabilities • HUMINT sources		X
Information Protect	• Centralized leadership • Unsecure communications	• Secure communications • Internet leaks	X	
Information Project	• Face-to-face • Inconsistent messages	• Radio broadcasts • Lack of credibility with populace	X	
Overall Information Advantage			X	

Figure 3-6. Information advantage work sheet

GENERATING INFORMATION SUPERIORITY

3-24. Once a thorough understanding of each side's capabilities and vulnerabilities is established and apparent advantages and disadvantages are determined, IO planners can begin generating options (COAs) to achieve information superiority.

3-25. Information superiority is an operational advantage derived from the ability to collect, process, and disseminate an uninterrupted flow of information while exploiting or denying an adversary's ability to do the same. Information superiority can be achieved by attacking the adversary force with information or shaping the information environment (or both). These attacks are directed at reducing any relative advantages the adversary has and exploiting its relative vulnerabilities in the information environment.

3-26. This duality of information operations—attacking the adversary and shaping the information environment—is analogous to fires and maneuver, where fires equate to attack of the adversary's ability to use information for C2 and as a weapon against friendly forces, and maneuver is an activity to seize and retain information nodes for the purpose of gaining a positional advantage in the information environment. To be effective, an information operation balances activities to attack the enemy force with those that shape the information environment. Through a combination of both, a military force seeks information superiority over the adversary. Figure 3-7 shows examples of information superiority.

Focus of IO	Cognitive Dimension	Physical Dimension
Adversary Force	• Slow decisionmaking • Reduce morale	• Misemployment of forces
Information Environment	• Change populace support	• Change populace behavior

Figure 3-7. Examples of information superiority

3-27. Once identified, information superiority becomes the purpose of the information operation, and as COAs are developed, they must be nested to the unit's main operation. To do this, IO planners determine the operational advantage (that is, information superiority) that will be sought in the information environment and ensure its purpose supports the purpose of the mission statement.

CONCEPT OF SUPPORT STATEMENTS AND SKETCHES

3-28. The IO concept of support describes how available forces will achieve information superiority. It states when and where information superiority needs to be achieved and describes how IO will support the operation and how IO capabilities will be employed. IO planners develop an IO concept of support for

each assigned mission or COA based on what the command's assets and resources can do to achieve the IO objectives. To build an IO concept of support, the IO planner develops the—
- Purpose of the information operation (information superiority).
- IO objectives or essential information operations tasks (EIOTs) that will create the effects in the information environment to achieve information superiority.
- Tasks to subordinate units and staff elements that assign specific actions that will achieve the IO objectives' desired effects.
- Target nominations. Certain IO tasks may result in the identification and nomination of targets.
- Request for support from higher HQ.
- Assessment plan to measure progress.

INFORMATION OPERATIONS OBJECTIVES

3-29. IO objectives describe the effects that will achieve information superiority. IO objectives do not stand alone, but support the commander's operational intent. As such, an IO objective is a statement of what IO will do to attack the adversary or shape the environment to achieve information superiority. For example, if information superiority for an operation is "prevent target from moving from Objective Black prior to attack," then IO objectives could be "disrupt adversary communications within Operational Area Blue to prevent early warning," "deceive adversary decisionmakers on Objective Black to prevent relocation of C2," or "influence local populace in Operational Area Blue to support friendly-force operations with preventing populace reporting of friendly-force activities."

3-30. For each mission or COA considered, IO planners develop IO objectives based on the tasks for IO identified during mission analysis. Depending upon the complexity or duration of the mission (for example, a tactical direct-action mission versus a long-term FID defense mission) there may be only one IO objective or there may be numerous IO objectives developed for each phase of the overall operation. Generally, regardless of the mission, no more than five objectives are planned for execution at any one time in the operation.

3-31. When possible, IO objectives should be observable (the desired effect is detectable), achievable (assets and time are available to accomplish the objective), and quantifiable (the desired effect can be measured). The effects describe a physical or cognitive condition either in the information environment (focus on information content and flow) or against adversary forces (focus on cognition and behavior). IO objectives should not specify ways or means (that is, IO capabilities).

3-32. There is no doctrinal format for an IO objective. One possible format uses target, action, purpose, effect:
- *Target* describes the object of the desired effect.
- *Action* describes the capability or cognitive function of the target.
- *Purpose* describes what will be accomplished for the friendly force.
- *Effect* describes the outcome (for example, destroy, degrade, disrupt, or deceive).

3-33. It is important that IO objectives are written in terms of effects, because it is the desired effect that focuses the activities (tasks) of IO capabilities. For IO, a proper effect falls into one of the three following categories:
- *Effects against the adversary.* IO effects against the adversary focus on the adversary's ability to collect, protect, and project information. An example IO objective is to **disrupt** (effect) **insurgent** (target) abilities to **conduct C2** (action) to **surprise** adversary forces in and around Village X (purpose).
- *Effects to shape the information environment.* IO effects shape information content and flow within the operational area's information environment. An example IO objective is to **influence** (effect) **local populace** (target) **perception** of the insurgents (action) to **increase reporting** of insurgent activity and locations to coalition forces (purpose).
- *Effects to protect friendly forces.* IO effects regarding friendly forces seek to prevent adversary interference with friendly abilities to collect, protect, and project information. An example IO

Planning Information Operations

objective is to **deny** (effect) **insurgent** (target) ability to **exploit** negative effects of friendly-force operations (action) to **prevent support** to adversary efforts (purpose).

Note. Figure 3-8 provides an example of directed effects.

3-34. Because it is impossible to anticipate all possible effects, terms other than those presented in this TC may be used to describe the desired effects for IO. Effects terms should describe a condition—not a task. Definitions may vary for the same effect based on the physical and cognitive nature of the effect and the target of the specific effect.

3-35. As IO objectives are developed, IO planners should consider the indications of success (MOEs) and how the indications will be collected. If adequate indications and collection means cannot be identified, the objective may have to be refined to produce measurable and detectable results. If an objective's MOE is focused on behavior or beliefs, planners must consider physical actions that are a result of the desired behavior, or belief, as an indicator.

Effects Against the Adversary		Effects to Shape the Environment	
Physical Effects	Cognitive Effects	Information Content	Information Flow
Destroy – Use lethal or nonlethal means to render adversary capabilities to collect, protect, or project information ineffective, unless reconstituted. **Degrade** – Use nonlethal or temporary means to reduce the adversary's effectiveness or efficiency to collect, protect, or project information. **Disrupt** – Interrupt the flow of information to and from the adversary and within the adversary organization. **Isolate** – Seal off an adversary from sources of support or contact with other adversarial elements.	**Deceive** – Mislead the adversary decision-makers, causing them to take specific actions or inactions that contribute to friendly-force mission accomplishment. **Influence** – Cause adversaries or others to behave in a manner favorable to friendly forces. **Isolate** – Prevent effective adversary decisionmaking by impeding the adversary's efforts to collect and project information.	**Destroy** – Use lethal or nonlethal means to render adversary information or information systems ineffective, unless reconstituted. **Degrade** – Use nonlethal or temporary means to reduce the effectiveness or efficiency of adversary message content. **Exploit** – Gain advantage of an adversary action that has negative effects on the populace. **Influence** – Cause adversaries or others to behave in a manner favorable to friendly forces.	**Degrade** – Use nonlethal or temporary means to reduce the effectiveness or efficiency of adversary communication methods with the populace. **Exploit** – Take advantage of gained access to a populace. **Isolate** – Prevent populace groups from communicating with each other. **Influence** – Cause information to move faster or slower, resulting in populace behavior that is favorable to friendly forces. **Disrupt** – Break or interrupt the flow of information between selected key-information nodes.
Effects to Protect Friendly Forces			
Deny – Withhold information about friendly-force capabilities and intentions that adversaries need to make effective and timely decisions. **Mitigate** – Reduce negative effects of friendly-force operations on the populace. **Neutralize** – Render an adversary's collection capability ineffective with regard to time, space, and purpose.			

Figure 3-8. Effects for information operations

ESSENTIAL INFORMATION OPERATIONS TASKS

3-36. At the tactical level, in a time-constrained environment, it is sometimes more straightforward to write IO objectives as EIOTs. EIOTs serve the same function as objectives; specifically, they focus the activities

(tasks) of the various IO capabilities. One difference is that EIOTs include the methods or means that will be used to perform the EIOT. A possible format for an EIOT uses task, purpose, method, effect:
- *Task* (for example, jam adversary's C2 communications).
- *Purpose* (for example, prevent coordinated efforts against friendly forces).
- *Method* (for example, EW EA-6B).
- *Effect* (for example, disrupt).

INFORMATION OPERATIONS TASKS TO SUBORDINATE UNITS AND STAFF ELEMENTS

3-37. Once IO objectives or EIOTs are written, IO planners develop tasks to subordinate units and staff elements that possess the IO capabilities needed to accomplish the IO objectives or EIOTs. Therefore, the aggregate execution of assigned tasks should achieve the effect of the IO objective or EIOT. Tasks for IO capabilities to subordinate units translate the broad concepts of the objectives and EIOTs into discreet actions. Tasks are often written as—
- *Task.* The task is the action to be performed and the location of the task (for example, prevent local populace interference in Village X).
- *Purpose.* The purpose is the reason why the task is assigned (for example, prevent civilian casualties).
- *Method.* The method describes what unit or capability will conduct the task (for example, MIS Team C121).

3-38. To develop tasks, IO planners should consider all available organic and supporting IO capabilities and resources that can help achieve each IO objective. As a matter of course, it is best to have representatives for each IO capability write their own tasks.

3-39. Similar to effects, tasks can be organized into three categories (Figure 3-9, page 3-13). These tasks are as follows:
- *Tasks against the adversary.* These tasks target adversary capabilities and vulnerabilities to collect, protect, and project information (as identified during the COG analysis). An example task is C*ounter insurgent propaganda to maintain populace support for capture/kill missions.*
- *Tasks to shape the information environment.* These tasks shape information content and movement by impacting the key nodes in each subinformation environment to influence local populace perceptions and behavior. An example task is *Engage religious leaders to stop inflammatory rhetoric.*
- *Tasks to protect friendly forces.* These tasks seek to protect friendly-force vulnerabilities in the information environment from adversary capabilities to collect and project information. An example task is *Detect intrusions into friendly-force information systems to prevent adversary collection of critical information.*

3-40. An IO planning work sheet (Figure 3-10, page 3-14) is a tool that can be used to develop an IO concept of support for input to COA development. One work sheet is filled out for each IO objective.

INFORMATION OPERATIONS CONCEPT OF SUPPORT

3-41. The IO concept of support is a word picture that explains how the information operation supports the operation from beginning to end, and how IO capabilities will be employed to provide information superiority. The IO concept requires defining information superiority for the operation. A well-written concept is concise and understandable. Although there is no doctrinally prescribed formula for an IO concept of support, leaders should consider the following:
- *Commander's intent for IO* describes what the commander wants IO to do to the adversary or to shape the information environment.
- *Information superiority* is described in the context of the operational situation and the command's mission; this should include the specific time and place for it to be achieved (should be linked to decisive points in the operation).

Planning Information Operations

- *General plan for IO* lists the IO objectives, tasks to be executed, capabilities that will execute, associated MOEs for the objectives, and collection methods that will be used for assessment.
- *Priority of support* designates which subordinate unit or element has the priority of IO assets and capabilities. *Restrictions on the employment of IO* lists prohibited and directed actions that affect the employment of IO.
- *General scheme for IO* uses doctrinal concepts and terms to explain how the IO objectives will be achieved, who will perform them (that is, the tasked units), and the sequencing of key tasks; it relates the key tasks to the achievement of information superiority.

Tasks Against the Adversary	Tasks to Shape the Environment
Counter – Diminish adversary information to correctly portray friendly intent and actions. **Demonstrate** – Show or reveal. MILDEC typically conducts tasks to demonstrate. **Deter** – Prevent action through the existence of a credible threat of unacceptable counteraction. MISO forces typically conduct tasks to deter. **Disseminate** – Spread or disperse. MISO and CMO typically conduct tasks to disseminate. **Jam** – Interfere with or prevent the clear reception of signals by electronic means. EW typically conducts tasks to jam. **Persuade** – Induce to believe something or convince. MISO forces typically conduct tasks to persuade. **Prevent** – Keep from happening or avert. OPSEC typically conduct tasks to prevent.	**Broadcast** – Transmit and make public by means of radio or TV. Typically MISO forces conduct tasks to broadcast. **Demonstrate** – Show or reveal. MILDEC typically conducts tasks to demonstrate. **Disseminate** – Spread or disperse. MISO and CMO typically conduct tasks to disseminate. **Engage** – Initiate contact to open dialogue with or communicate a message to a target. MISO and CMO typically conduct tasks to engage; however, any friendly-force asset with access to the target has potential to conduct face-to-face engagements. **Inform** – Provide information or educate a specific target audience. MISO, PA, and CMO typically conduct tasks to engage; however, any friendly-force asset with access to the target has potential to inform through face-to-face engagements. **Persuade** – Induce to believe something or convince. MISO forces typically conduct tasks to persuade. **Publicize** – Bring to the attention of the public. PA typically conducts tasks to publicize.
Tasks to Defend Friendly Forces	
Detect – Discover or discern the existence, presence, or fact of an intrusion into information systems. IA, CI, and EW typically conduct tasks to detect. **Protect** – Guard against espionage or capture of sensitive equipment or information. OPSEC, IA, CNO, physical security, EW, and CI typically conduct tasks to protect. **Respond** – React quickly and appropriately to an adversary attack or intrusion in the information environment. All IO capabilities have potential to respond, depending on the specific incident. **Restore** – Bring information systems or conditions in the information environment back to their original state. IA typically conducts tasks to restore.	

Figure 3-9. Tasks for information operations capabilities

Chapter 3

COA: Conduct raid at Objective LIMA to remove insurgents and return control of the area and populace to the existing government. **IO Objective:** Disrupt communications by jamming insurgent communications. This will prevent coordinated efforts against friendly forces.		
EW Tasks: Provide electronic jamming of mission command communications used by the insurgents.	**Information Operations Targets:** Insurgent command and communications nodes.	**Protect Assets:** Protection of friendly communications is essential.
MISO Tasks: None		
OPSEC Tasks: Protection of the essential elements of friendly information is imperative to ensure success of electronic warfare plan.		
MILDEC Tasks: None	**MOEs:** Inability of insurgent forces to send early warning and inability to communicate during the mission.	
CMO Tasks: None	**Intelligence Requirements:** Insurgent key personnel and frequencies used by key communicators in the area of operations.	
PA Tasks: None		
Other Tasks: None	**Coordination:** Coordinate with the spectrum manager and adjacent and higher headquarters.	

Figure 3-10. Example of an information operations planning work sheet

INFORMATION OPERATIONS CONCEPT OF SUPPORT SKETCH

3-42. The IO concept of support sketch is a visual graphic (Figure 3-11, page 3-15) of the information operation. It is the product used to brief the commander and staff on what IO capabilities will do during the mission. The format or medium used for the sketch is not as important as ensuring the correct elements of information are presented, that the sketch shows that the information operation is synchronized with other operations, and that it clearly depicts the synchronization of the IO capabilities involved with the operation. The sketch should answer the following questions:

- *Who?* What capabilities will be employed to perform the IO tasks?
- *What?* What operational advantage is provided by the information operation (that is, information superiority)? What objectives (or EIOTs) must be achieved and what are the required tasks to IO capabilities?

Planning Information Operations

- *When?* What time during the operation will tasks be performed?
- *Where?* Where in the operating area will the IO tasks be performed?
- *Why?* What is the purpose of each IO task?

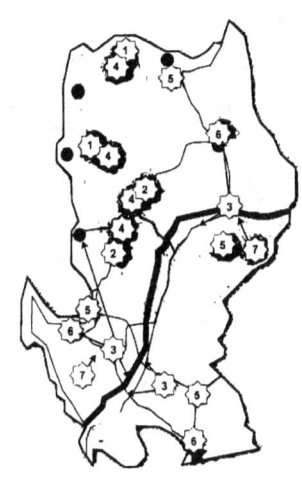

Information Superiority: Decrease opposition to friendly-force operations IO objectives:
- Influence insurgents to disarm and disband, IO task: Increase populace confidence in the established government.
- Exploit insurgent intimidation tactics against the populace, IO task: Increase populace support to friendly-force objectives.

IO Element Tasks:
MISO
1. Disseminate leaflet to suspected insurgent areas, IO task: Decrease insurgent's will to fight.
2. Employ MIS teams to disseminate handbills to known insurgent areas, IO task: Increase populace participation in humanitarian assistance distribution.
3. Employ MIS teams in direct support of maneuver units, IO task: Prevent populace interference.
4. Inform populace of recent insurgent related atrocities, IO task: Prevent insurgency recruitment.

CA
5. Distribute humanitarian assistance to populace in insurgent areas, IO task: Reduce populace support to insurgency.
6. Distribute humanitarian assistance to dislocated civilians camps, IO task: Increase support to friendly-force operations.

PA
7. Publicize peacekeeper's role in humanitarian assistance projects, IO task: Prevent effective adversary propaganda.

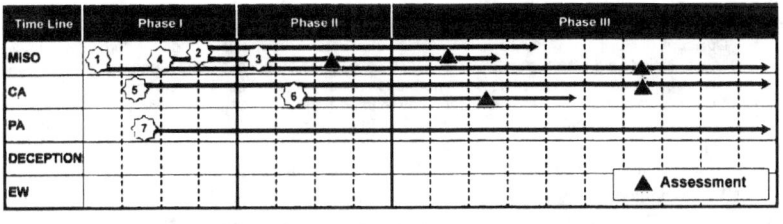

Figure 3-11. Example information operations concept of support sketch

ORDERS PRODUCTION

3-43. Plans and orders are as detailed as time permits. The size of these documents depends on the command and mission; they can run the gamut—from a series of overlays with written comments to voluminous documents of hundreds of pages. Regardless of the format used, an order must be clear, concise, timely, and useful to the implementing commands and units.

3-44. The IO annex (Army orders format) or appendix (joint orders format) describes the complete IO mission and how IO will gain information superiority in support of the scheme of the maneuver. This approach places a lesser emphasis on individual IO assets and capabilities and greater emphasis on the aggregate IO effects needed to achieve information superiority. The IO staff must be careful to not let the requirement to develop and explain IO capabilities contribution to the operation overwhelm the primary purposes of the IO annex, which are to—
- Provide operational details on the information operation.
- Focus element and unit tasks on achieving specific effects in the information environment.
- Provide the information needed to assess the information operation.

3-45. There are two basic formats for an IO annex: a five-paragraph (Figure 3-12, pages 3-16 and 3-17) and a matrix annex. The five-paragraph annex is used when sufficient planning time is available. The matrix annex is used when time is limited or when directed by the J-3/G-3/S-3 or unit SOP.

ANNEX P (INFORMATION OPERATIONS) TO OPERATIONS ORDER NO ##

1. **SITUATION.**
   a. **AO.** Describe the information environment's subenvironments. Identify significant characteristics (for example, terrain, weather, populace, civilian information infrastructure, civilian population, and third-party organizations). State the aggregate impact on adversary and friendly operations. Identify aspects of the information environment, to include key information nodes that favor adversary and friendly operations.
   b. **Adversary Operations in the Information Environment.** Describe how, when, where, and why adversary forces will operate in the information environment. Describe likely objectives and activities and how information capabilities will be employed. Identify adversary capabilities and vulnerabilities in the information environment in terms of information collection, protection, and projection.
   c. **Friendly Capabilities and Vulnerabilities in the Information Environment.** Identify friendly-force capabilities to shape the information environment and attack adversary forces with information.
   d. **Civil Considerations.** Identify key people, groups, and organizations that operate in the information environment and will affect friendly and adversary forces' operations. Describe likely objectives and activities in the information environment.
   e. **Attachments and Detachments.** List organic and supporting assets that are available to execute the information operation.
2. **MISSION.** State the unit mission.
3. **EXECUTION.**
   a. **Concept of Support.** Describe how IO will be conducted and who will perform it from beginning to end, to include adversary capabilities and vulnerabilities to be attacked and friendly critical vulnerabilities to be protected. Define information superiority (that is, the operational advantage derived from operating in the information environment) and explain how and when IO will achieve it. Include IO effects (that is, objectives or EIOTs), sequencing of key tasks, and IO capacities priorities by phase.
   b. **Assessment.** Describe the assessment plan for the information operation.
   c. **Tasks to Subordinate Units.** List subordinate units and assigned IO tasks.
   d. **Coordinating Instructions.** List IO instructions common to two or more units. State any rules of engagement applicable to IO capability. List constraints not contained in the concept of support.
4. **SERVICE SUPPORT.** Identify requirements for support pertaining to IO as a whole. Identify service support to individual IO elements in their respective appendixes or annexes.
5. **COMMAND AND SIGNAL.** Significant command and signal information related to IO not covered in the base order. Include arrangements needed to exchange information among IO capabilities.

ACKNOWLEDGE: (If distributed separately from base plan/order)

Figure 3-12. Format for a five-paragraph information operations annex (Army orders format)

[Authenticator's last name]
[Authenticator's rank]
APPENDIXES:
1. OPSEC
2. MISO
3. MILDEC
4. EW
5. IO Execution Matrix

Figure 3-12. Format for a five-paragraph information operations annex (Army orders format) (continued)

3-46. Typically, at the tactical level, the information operation can be adequately described on a matrix order format (Figure 3-13). When combined with a copy of the IO concept of support sketch, most IO capabilities can understand and execute accordingly. There is no specific format for an execution matrix. Figures 3-14 and 3-15, pages 3-18 and 3-19, are two examples.

Enemy Situation:	Friendly Situation:
See Appendix 1 to Annex B	XXI Corps EC-130H, EC-130E, EA-6B, F-16CJ (HARM), AC-130 (Specter)

Mission:	Information Superiority:
Prevent preemption of air assault; influence local population to not interfere in and around the objective; shape the information environment to establish order and provide basic services.	Dominance of the information environment which permits mission success without effective opposition and minimal civil interference.

Concept of Support:
Prevent preemption of the air assault and minimize civil interference in and around the objective by destroying, degrading, disrupting, and exploiting adversary mission command and fire support systems; deceiving adversary decisionmakers; destroying, degrading, disrupting, and deceiving enemy information systems; denying adversary decisionmakers information about XXI Corps intentions and capabilities; protecting mission command and information systems.

IO Objectives/Tasks:	Assessment:
Prevent compromise of the operation; protect XXI Corps mission command; disrupt 109th Division air defense and targeting systems during critical periods of the operation; minimize civilian interference.	

Coordinating Instructions:
XXI Corps: Contact counterparts to coordinate and synchronize efforts to identify suspected SPF locations.

Service Support:
No change.

Command and Signal:
XXI Corps IO cell is located in the Main CP.

Appendixes:
Appendix 1 (OPSEC), Appendix 3 (EW), Appendix 4 (IO Execution Matrix).

Figure 3-13. Example of a format for a matrix information operations annex (Army orders format)

Capability	Phase I	Phase II	Phase III	Phase IV
EW	Monitor signals of interest. Electronic protection for personnel and equipment.	Electronic attack to disrupt enemy communications. Electronic protection for personnel and equipment.	N/A	N/A
MISO	Broadcast harassment messages for enemy. Broadcast noninterference messages for local populace.	N/A	Broadcast via mobile radio to keep population informed on mission.	Broadcast on mission success. Coordinate with combat camera crews for post-mission propaganda and counter-propaganda.
OPSEC	Determine essential elements of friendly information for mission.	Implement measures to protect essential elements of friendly information to protect movement routes, mission command, and objective.	N/A	N/A
CNO	Maintain computer network defense to protect friendly communications and information.	Maintain computer network defense to protect friendly communications and information.	Maintain computer network defense to protect friendly communications and information.	Maintain computer network defense to protect friendly communications and information.
MILDEC	N/A	N/A	N/A	N/A
CMO	Prepare Commander's Emergency Response Program paperwork for funds disbursement. Coordinate with provincial reconstruction team.	N/A	N/A	Assist personnel returning to villages. Assess small-scale immediate projects.
PA	Prepare press releases.	N/A	N/A	Send press releases. Control local and national media.
COMCAM	Document operation.	Document operation.	Document operation.	Document operation.
HN Forces	Take lead in all lethal actions on the objective.	Take lead in all lethal actions on the objective.	Take lead in all lethal actions on the objective.	Take lead in all lethal actions on the objective.
Other	N/A	N/A	N/A	N/A

Figure 3-14. Example 1 of an information operations execution matrix

Planning Information Operations

Tasked Unit or System	IO Task	Time on Target or Time of Effect	Location	Remarks
EA6B	EW-01	H–1 through H-hour	Throughout area of operations	Successful if enemy is unable to send early warning
TPT	PSY-01	H–24 and continue	Objective LION	Successful if no civilian interference
95th Civil Affairs Brigade	CMO-01	H–24 though H-hour	Through area of operations	N/A
Special Instructions: None				

Figure 3-15. Example 2 of an information operations execution matrix

CONSIDERATIONS

3-47. IO planning can be initiated at the SFODA typically by the attached MIS element as the only information capability at that echelon, and finalized at the SOTF. Typically, an SFODA will not have access to all of the IO capabilities. When developing a concept of operations, planners must consider and include the applicable capabilities in Figure 3-14, page 3-18. The SOTF IO planner will coordinate for assets and synchronize and deconflict the effects. It is important for the SFODBs and SFODAs to understand the JSOTF/SOTF IO plans to ensure the higher HQ intent is nested within their plans.

CONSEQUENCE MANAGEMENT

3-48. When planning, the SFODAs should consider what actions to take if the operation does not go as planned. In the event that an operation does go awry, it is important to understand the information environment, to include the influential leaders that SFODAs and their FID partnered units should engage to assist in getting the appropriate information and messages to the populace. Also, it is important to note what media outlets are available to assist with getting ahead of the news cycle. Key consideration is that the first voice is the loudest. If the SFODA or the HN provides the facts of an operation or event in a timely manner, the adversary will be forced into defense and will have to react and counter the information as they fight to influence the population.

This page intentionally left blank.

Chapter 4
Execution of Information Operations

This chapter focuses on the key staff tasks IO planners must accomplish during execution of an information operation. Once execution begins, IO planners monitor the adversary and friendly situations, track IO task accomplishment, determine the accomplishment of the IO objectives and tasks, and detect and track any unintended consequences. Three staff tasks critical to execution include the following:

- *Monitor.* Planners must maintain situational awareness and monitor the progress of operations to determine if the operation is going according to plan.
- *Evaluate.* Planners analyze the progress of the information operation, the status of the adversary, and the effects in the information environment to determine if there are variances from the plan and the significance of the variances.
- *Adjust.* Planners estimate the effectiveness of task execution and the effectiveness of IO on the adversary, the local populace, and friendly operations. If an unexpected incident occurs, IO planners—in coordination with current operations staff—coordinate with subordinate units and staff elements to develop an appropriate task for that incident.

MONITORING

4-1. The key to monitoring is the collection of information critical to execution of the information operation. The first step is determining what information is needed to evaluate and adjust the information operation. Two sources for deriving the information are the CCIR and the J-2/G-2/S-2 decision support template. From these sources, IO planners can determine their own information requirements; notably, intelligence requirements and FFIRs that will help guide the information collection effort.

4-2. Next, IO planners monitor both the command's overall operation and the IO tasks and activities as spelled out in the operation order, IO appendixes, annexes, and execution matrices. Then, operations reports and intelligence summaries are reviewed for IO-relevant information and paired against the IO objectives to evaluate progress of the information operation. If necessary, requests for information are submitted and tracked for clarification or additional information.

EVALUATING

4-3. The purpose of the assessment is to judge success or progress of the information operation. Progress is determined by analyzing relevant information and intelligence from unit operations and intelligence reports. The information is then applied against current IO objectives to determine whether the desired effects are being achieved. However the assessment is conducted, planners should consider the following principles:

- The assessment should lead to recommendations to the commander to continue, end, or change the operation.
- The assessment must detect situation changes quickly enough for commanders to respond effectively.

Chapter 4

- A balanced assessment considers changes in both the friendly and adversary forces' information environment.
- The assessment of an information operation should focus on collective rather than individual tasks and targets, because changes in the information environment or adversary force usually are not the result of any single task or target.

4-4. Assessing IO can appear complex and difficult, but it need not be burdensome if IO planners use a simple methodology to assess the information operation. In principle, assessing an operation consists of evaluating the operations against measures of performance (MOPs) and MOEs.

Note. IO planners must be careful not to over assess by becoming bogged down in formal assessment procedures for numerous tasks and effects, or to overwhelm subordinate units or staff elements with requirements for numerous reports, questions, and information requirements.

4-5. MOPs measure friendly actions in terms of task accomplishment and performance. IO cannot generate effects if the planned tasks are not successfully executed. As such, assessment should account for task execution. Because task completion affects execution as well as assessment, it is important that the IO capability representatives and subordinate units report accomplishment of their respective tasks. MOPs are not measures of success—they gauge task completion and do not measure effect success or failure.

4-6. MOEs are used to measure the results achieved in the overall mission and execution of IO objectives. More practically, MOEs determine if a desired condition or outcome is in place (that is, effect), even if not directly caused by planned military action. Because IO objectives are written to articulate a specific condition or state in the operational environment, most MOEs are crafted and used to measure the effects generated by those tasks collectively executed to achieve each IO objective.

4-7. An assessment plan is normally developed as part of the planning process. For complex or long-term operations, it may be necessary to form an assessment working group to produce the information required to assess the information operation. Attendees to the working group may include representatives from the J-2/G-2/S-2; J-3/G-3/S-3; plans directorate of a joint staff (J-5); assistant chief of staff, plans staff section (G-5); MISO; EW; CA; and PA. During combat operations, the combat assessment board may supplant the assessment working group.

Note. One of the critical factors in a successful relief-in-place or transfer of authority between IO planners is the passing on of all historical IO data and ongoing assessment plans.

4-8. There is no standardized or doctrinal assessment process. In the absence of a doctrinal process, IO planners must develop their own methodology to guide assessment. Based on field experience, the following process is a logical approach to assessing an information operation:

- *Develop assessment criteria.* The first step in the assessment process is to develop the assessment criteria. This involves developing the items that support the MOE, indicators, and MOPs. One or more MOE is normally developed for each IO objective. Each MOE should clearly articulate the desired condition (effect) or end state that supports the associated IO objective. For example, for an IO objective to "reduce popular support for insurgents," an MOE could be "level of popular support to insurgents." Next, indicators are developed for each MOE. Multiple MOEs should be developed and used to determine if the IO objective is being achieved. Indicators for the MOE "increased reporting of insurgent activities" could include increased tips-line reporting, increased tips to patrols, and increased walk-in tips. MOPs assess task accomplishment. Example MOPs include the number of face-to-face interviews conducted, the number of handbills disseminated, and the number of radio broadcasts (all observable and measurable activities).
- *Define the measures.* Once MOEs, indicators, and MOPs are developed, IO planners establish a foundation for comparison and analysis (also known as *benchmarking*). Benchmarking determines the current state of MOEs and supporting indicators.

- *Collect and analyze data.* The next step is to identify sources of data needed to assess the MOEs, indicators, and MOPs. Data collection requirements should be kept as simple as possible. When possible, standard operations and intelligence reports are used as the means to collect the data. Figure 4-1 shows some data collection sources.
- *Provide recommendations.* Assessment should yield changes to execution (based on MOPs), changes to desired effects (based on MOEs), and changes to resource allocation. Recommendations based on the assessment should provide the commander with the bottom line, a recommended way ahead, and any issues requiring the commander's involvement.

	Data Collection Sources
Internal	Situation reports, intelligence summaries, current operations data, and other command reports.
Organic	HUMINT, CA, MISO, and PAO reports; significant acts database; subordinate unit assessments.
External	Other government agencies, international organizations, polling and populace surveys, media analysis, OSINT.

Figure 4-1. Sources of data collection

4-9. As the data is collected, it should be analyzed against the established indicators to establish the benchmarks. Follow-on data collection periods and assessments establish changes to the indicators and MOEs. The sum of indicators then provides the assessment for each MOE. Figure 4-2 shows an example of an assessment graphic.

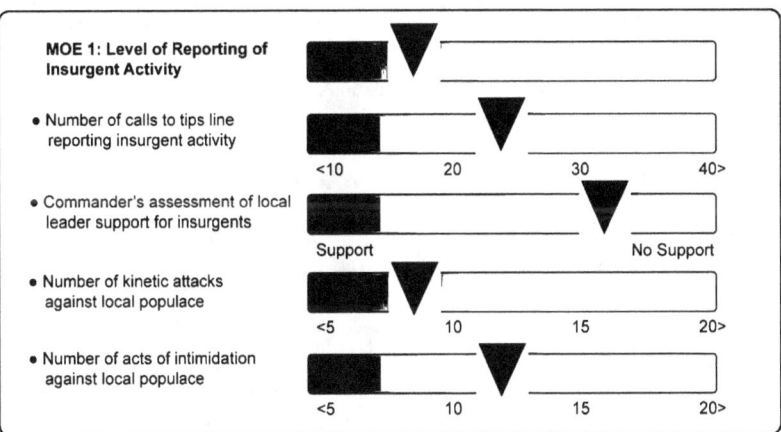

Figure 4-2. Example of an assessment graphic

ADJUSTING

4-10. The ways that IO planners can adjust the information operation in response to events in the information environment are battle drills, the IO working group, and crisis-action teams. In a best-case scenario, IO planning accounts for all possibilities and sets conditions for further operations. The information environment is never static, and planning for consequence management through battle drills and rehearsed crisis-action teams is critical to staying ahead of the adversary's information cycle. By the same token, proactive planning of synchronized IO efforts helps preclude reactive IO responses

Chapter 4

for example, facilitating more timely and better prepared MISO and PA products and preplanned key-leader engagements.

BATTLE DRILLS

4-11. Staff battle drills are planning aids designed to speed response to crisis situations that occur during the conduct of a mission. For IO, quick responses to the adversary's actions and events in the operational area are necessary to beat the adversary in the information environment and ultimately achieve information superiority.

4-12. Battle drills are developed during the planning process; however, they are not complete and final COAs. Rather, battle drills are predeveloped concepts that anticipate crises. Once a crisis occurs, the battle drill (that is, the COA) can be quickly adjusted to address the realities of the situation at hand.

4-13. A military operation can be thought of as a series of events, planned and unplanned, that force both friendly and enemy forces to react to a changing situation. Some of these events, referred to as critical events, are keys to mission success of friendly or enemy forces. Critical events—

- Can create both intended and unintended effects and may be brought on by friendly, adversary, or third-party actions.
- Can be either negative or positive. The staff can develop drills that react to either type. For negative critical events, a battle drill should mitigate the impact of the event on the populace and friendly forces. For positive critical events, a battle drill should exploit the event to maximize the impact on the populace and adversary forces.
- Can be triggers or cues for the staff to initiate a battle drill.

4-14. An IO battle drill is a generic concept of support that addresses a friendly-force IO response to a critical event that may occur during execution of the operation. There is no established format for battle drills, though it should be recognizable to the staff and mirror existing products. Development of battle drills does not follow an established guide, but rather, they are developed to suit specific missions and potential branches and sequels of missions. Each battle drill should—

- Identify critical events.
- Define information superiority.
- Develop IO concept of support.
- Determine tasks and targets.

The information contained in a battle drill is not a final and complete plan, but rather a concept that must be refined to the realities of the situation at hand. Depending on the battle drill, productions of approved MISO products (such as radio scripts or other products) may be appropriate.

Identify Critical Events

4-15. Planners determine what critical events may result from friendly, adversary, or third-party action. During an upcoming operation, planners focus on events that will either occur in or affect the information environment and are significant enough to affect the command's mission. The following list provides some examples of critical events:

- Civilian collateral damage.
- Civilian casualties.
- Fratricide incidents.
- Populace interference with friendly-force operations (for example, civil demonstrations).
- Quick-reaction force deployment.
- Adversary or friendly forces violation of law of land warfare (for example, atrocities against civilians, mass-grave discovery).
- Environmental incident (for example, hazardous-material spill).
- Propaganda directed against friendly forces.
- EEFI or any other sensitive or classified information disclosure.

Execution of Information Operations

Define Information Superiority

4-16. Battle drills are designed to respond to a specific situation. Therefore, the situation must be sufficiently defined so the IO planners can adjust the battle drill's concept to compensate for the differences between the planned and actual situation. For IO, this means defining information superiority for each battle drill. Information superiority is the operational advantage provided to the commander through the control and management of information content and flow in the AO. Examples of information superiority for mitigation and exploitation battle drills are as follows:

- A mitigation battle drill:
  - *Event.* Disclosure of EEFI or classified information.
  - *Target.* Adversary.
  - *Information superiority.* Adversary decisionmakers are unable to take advantage of sensitive information about the friendly force.
- An exploitation battle drill:
  - *Event.* Destruction of key infrastructure by adversary.
  - *Target.* Populace.
  - *Information superiority.* Populace does not support the actions of enemy forces.

Develop Information Operations Concept of Support

4-17. The concept of support is a concise and easily understandable word picture describing how IO capabilities may be employed and what staff coordination must be conducted to employ the capabilities. The concept must be integrated with the overall operation, when applicable. How much information is known when the battle drill is created determines the level of detail. The IO concept of support should include the following:

- *Assumptions.* Planners list information accepted as true in the absence of facts at the time the battle drill is developed. Planners periodically review and update the battle drill by validating the assumptions.
- *Information superiority.* Planners determine and then describe the operational advantage IO will provide.
- *General scheme for IO.* Planners use doctrinal concepts and terms to explain how IO will achieve information superiority, listing any IO objectives, EIOTs, and who will perform each key task at what time.

Note. Where the tasks are performed is determined once the battle drill is put into action.

- *Priority of support.* Planners designate which subordinate unit or element has priority use of IO capabilities.
- *Constraints on IO.* Planners list prohibited and directed actions that are expected to affect the information operation, paying particular attention to information content and flow (for example, no jamming in urban areas).

Determine Tasks and Targets

4-18. Leaders develop tasks, purpose, methods, and means, and, if appropriate, targets for each participating IO capability. A purpose for each task is included to explain each capability's part in the operation. If appropriate, general target sets are identified for each tasked element or capability. All IO-relevant capabilities—maneuver units and those staff entities that may have important roles in responding to the battle drill event—are considered. A purpose for each task is included to maximize asset initiative. Supporting elements develop MOP for their assigned tasks.

> *Note.* The formal battle drills are developed and coordinated at the SOTF level. The SOTF IO planner will use the 7-day mission tracker to ensure they are ready to provide support upon start of mission.

EXAMPLE STAFF BATTLE DRILL

4-19. There are several different battle drill formats currently in use. Figure 4-3 provides a sample format that has worked well in the field. Leaders should modify the format as needed to fit the command's needs and situation. Figure 4-4, page 4-7, provides an abbreviated staff battle drill.

SITUATION: Insurgent forces attack friendly forces, a friendly third-party organization, or an opposing faction (for example, a bombing, shooting, or mortar attack).

ASSUMPTIONS: The insurgent attack does not cause significant friendly causalities.

LIKELY FRIENDY ACTION: A response force is deployed to secure the site, and find and destroy the insurgent force. Security operations are conducted in and around the area of attack. If necessary, force protection measures are increased.

IO CONCEPT: The purpose of this IO is to gain populace support for counterinsurgency activities and identify hidden insurgent cells for targeting. IO capabilities provide direct support to the response force. MIS teams disseminate print products to the populace near the attack site. Unit leaders, MIS teams, and CA teams engage local leaders to gain support for friendly operations. PAO issues a press release to explain the command's position and counter misinformation concerning the situation. **Restrictions:** MISO products must conform to and support approved programs. **MOE:** Increased reporting of insurgent activity by populace.

Capability	Key Tasks	Purpose	Method	Target
MISO	Disseminate print products and radio broadcasts to the populace of villages in and around the attack site.	Identify hidden insurgent cells. Reduce populace support for insurgent forces and activities.	Handbills and posters. Contact radio.	Local populace. Insurgent fence-sitters.
CA	Engage local leaders.	Gain support for counterinsurgency activities.	Face-to-face.	Civil leaders.

Figure 4-3. Battle drill format for insurgent-related violence

INFORMATION OPERATIONS WORKING GROUP

4-20. An information operations working group (IOWG) consists of staff representatives who meet to coordinate and provide recommendations for the planning, execution, and assessment of IO. The IOWG also is used to synchronize the contributions of the IO capabilities. Participation in the IOWG is typically a mix of staff representatives and subject-matter experts.

> *Note.* IOWGs are formed at the SOTF and JSOTF levels.

4-21. The frequency of IOWG meetings depends on the situation and echelon. The working group may gather daily, weekly, or monthly depending on the situation, echelon, and time available. The formality of the IOWG also varies by echelon. For purposes of organization and focus, even the simplest IOWG should have an agenda. The composition of the IOWG is tailored to the agenda. Representatives from every staff need not attend every IOWG. Participants are selected because they either represent a critical element or capability or because they have expertise that is critical to the information operation. Typical attendees include the following:

- IO planners.
- MIS representatives.
- EW representatives.
- OPSEC representatives.
- COMCAM representatives.
- CA representatives.
- PA representatives.
- MILDEC representative.
- Representatives from the J-2/G-2/S-2.
- Representatives from the J-3/G-3/S-3, effects cell.
- Special technical operation planners.
- Fire support officer.
- Others, as required.

Note. Appendix A, pages A-1 through A-6, provides detailed information on IOWG.

---

1. **Situation:** React to collateral damage resulting from coalition-force action.
2. **Information Superiority:** Preempt adversary propaganda and negative media reporting.
3. **Immediate (on-site):**
   - Notify commander.
   - Document the scene (for example, COMCAM photos).
   - Conduct on-site key-leader engagement to determine facts and conduct initial mitigation.
4. **Within 2 Hours:**
   - Notify operational-area owner.
   - Notify local-government officials.
   - IO coordinates and synchronizes a public statement of the facts for broadcast by local print, radio, and TV media.
5. **Within 24 Hours:**
   - Conduct key-leader engagements with local elders using HN partner-unit commanders, coalition commanders, and local-government officials.
   - Assess damage for possible CMO projects.
6. **After 24 Hours:**
   - Coordinate for follow-up media coverage and key-leader engagement by operational-area owner.
   - Compensate family (if appropriate) and conduct CMO activities.

Figure 4-4. Abbreviated staff battle drill

CRISIS-ACTION TEAM

4-22. For significant operational matters, a crisis-action team may be activated. The crisis-action team consists of key members of the staff, to include the IO officer. When activated, the crisis-action team plans and rehearses the command's reaction to the event, and then issues a fragmentary order. To be a viable participant, IO planners should develop response options to support the crisis-action team planning process. When possible, IO planners use battle drills as the basis for adjustments to the information operation and tasks to the IO capabilities.

Chapter 4

REPORTING

4-23. Significant events and friendly- and adversary-force activity in the information environment should be routinely reported to the J-2/G-2/S-2, J-3/G-3/S-3, and, as appropriate, the higher HQ' IO staff. The guiding principles for reports are to—
- Keep the report as simple and as short as possible.
- Include only that information which feeds a planning, assessment, or reporting requirement.

4-24. At the JSOTF, it may be useful to develop an IO intelligence summary or IO operation summary. Depending on the mission and tempo of the operation, these reports may be daily, weekly, or monthly products. The IO situation report is an event-driven report that provides basic information on significant activity in the information environment as it occurs.

INFORMATION OPERATIONS INTELLIGENCE SUMMARY

4-25. An intelligence summary is provided to subordinate commands, interested staff elements, and higher HQ. The primary focus of the report is to capture significant events in the information environment (focused on the IO planners' intelligence requirements) and assess their impact on friendly- and adversary-force operations.

INFORMATION OPERATIONS OPERATION SUMMARY

4-26. Subordinate units provide operation summary reports on the status of IO in their respective AOs. The primary focus of the report is assessment. The report provides recent significant activities, current and planned operations, capability status, and assessment of IO objectives, key tasks, and engagements.

INFORMATION OPERATIONS SITUATION REPORT

4-27. The purpose of the IO situation report is to provide an update since the last reporting period. An IO situation report is event-driven by significant changes to the characteristics of the subinformation environments with regard to information content and flow, or modifications to adversary actions that address the characteristics. The information in the situation report usually consists of the 5Ws (who, what, when, where, and why). Situation reports are rendered as needed.

4-28. At the SOTF level and below, an IO summary should be produced. The IO summary should include significant events in the information environment, focusing on—
- IO planners' intelligence requirements and assessments of their impact on friendly- and adversary-force operations.
- Recent significant activities, current and planned operations, capability status, and activities and assessment of IO objectives and key tasks.
- Engagements and significant changes to the characteristics of the subinformation environments with regard to information content and flow.
- Modifications to adversary actions that address the characteristics.

Chapter 5
Intelligence Support to Information Operations

Intelligence is the product resulting from the collection, processing, and integration of information and knowledge about adversaries and their networks obtained through observation, investigation, analysis, or understanding. IO planning and execution rely on the existing intelligence capabilities of the command to provide support. IO significantly increase the demand for intelligence and require detailed analysis of the information environment and the adversary's use of the information environment.

Intelligence support to IO is not solely an intelligence-community task. The intelligence staff is responsible for coordinating and overseeing all command intelligence; however, each staff section and element involved in planning and execution has a responsibility to assist in this task. Thus, IO planners should work closely with intelligence personnel throughout the intelligence cycle to ensure effective intelligence support, but they must also conduct their own research and analysis.

Intelligence support to IO is continuous and requires long lead times. The intelligence necessary to affect the perceptions and decisionmaking of adversaries or other audiences often requires that specific sources and methods be positioned and employed to collect the information and conduct the analyses needed for the information operation. The challenge is to get the right information and intelligence at the right time.

As in other intelligence activities, analysts should be careful not to describe or portray the adversary's actions in the information environment as a mirror image of U.S. IO concepts, doctrine, and TTP. Culturally, the adversary is unlikely to think or act as the United States does.

The key terms used in this chapter are defined below:
- *Information requirement.* Information requirements are information elements required for planning, executing, and assessing operations.
- *Intelligence requirement.* An intelligence requirement is a requirement for the intelligence system to fill a gap in the commander's and staff's knowledge or understanding of the operational environment or threat.
- *Priority intelligence requirement.* The commander designates PIRs. PIRs are requirements associated with a decision that affects mission accomplishment. Information requirements not designated by the commander as PIRs become intelligence requirements.
- *Intelligence estimate.* An intelligence estimate is an appraisal of available intelligence relating to a specific situation or condition with a view to determining the COAs open to the enemy or adversary and the order of probability of their adoption.
- *Intelligence preparation of the operational environment.* IPOE is an analytical methodology employed to reduce uncertainties concerning the enemy,

environment, and terrain for all operations. IPOE builds an extensive database of potential areas where units may be required to operate. The database is analyzed in detail to determine the impact of the enemy, environment, and terrain on operations and then presents the analysis in a graphic form. IPOE is a continuing process. In joint doctrine, IPOE is referred to as joint intelligence preparation of the operational environment.

INFORMATION OPERATIONS AND THE INTELLIGENCE CYCLE

5-1. All intelligence for the command and staff, to include that needed for IO, is produced as part of the intelligence cycle. By working closely with the J-2/G-2/S-2 during the intelligence cycle, IO planners can minimize intelligence gaps and maximize available intelligence and collection assets to develop a reasonably accurate understanding of the information environment and a representative and reliable model of adversary operations in the information environment. To integrate into the intelligence cycle, IO planners—

- Identify intelligence gaps (IO-specific) concerning the information environment and adversary operations in the information environment, develop PIRs, and submit requests for information to fill the gaps.
- Become familiar with available collection assets, capabilities, and support relationships (direct support or general support). Planners determine time requirements for each collection asset and consider the capabilities and limitations of the assets that will perform the mission.
- Coordinate with the collection manager to ensure information requirements for IO are considered for inclusion as collection tasks.
- Establish relationships with key intelligence personnel. Planners should not go directly to an analyst without awareness or concurrence of J-2/G-2/S-2 leadership.
- Vet all intelligence products developed from reachback support and other external sources through the J-2/G-2/S-2 to avoid disconnected analysis.
- Provide feedback on the quality of intelligence provided and its usefulness to facilitate refinement.
- Assess the intelligence support that is provided to improve the working relationship with the intelligence staff while providing feedback to the intelligence analyst for improvements.

INTELLIGENCE "PUSH" AND "PULL"

5-2. Intelligence is disseminated by either the "push" or "pull" principle. For "push," IO planners must coordinate with the J-2/G-2/S-2 staff to get access to the dissemination means that have IO-pertinent products. This is accomplished by working with the intelligence analysts to get IO-specific information requirements injected into the collection cycle, nominating PIRs for either the information environment or adversary actions in the information environment, and coordinating with higher HQ IO staffs to get routine access to their intelligence products. To "pull" intelligence from the J-2/G-2/S-2 staff, IO planners should coordinate for access to those assets and systems that have IO-relevant information and intelligence, attend J-2/G-2/S-2 staff updates and fusion meetings, and coordinate with troop units and specialized teams (SFODAs, MIS teams, CA teams, and so on) to collect and report collateral information that is relevant to IO.

5-3. OSINT is an often overlooked way to get information and intelligence. Much useful information about the populace and media is available from public sources. This information often addresses the IO's information and intelligence requirements. Like other aspects of planning, IO planners must be prepared to conduct their own OSINT gathering.

REQUESTS FOR INFORMATION

5-4. Intelligence production is requirements-driven. Requests for information are used to request specific information and intelligence. Each command has its own requests for information format and procedures; however, the following rules should be observed when developing requests for information:

- *Conduct initial research.* Units should try to find the information or intelligence on their own, using requests for information to get information that is not readily available. Sources already checked should be listed so the intelligence analyst does not waste time working with materials and products that do not have the requested information.
- *Clearly state the requirement.* Units should describe—as specifically as possible—what information is needed. Language and terms associated solely with IO should be avoided, as should requests for a particular type of intelligence (for example, SIGINT or HUMINT). Requests for information should be restricted to one question.
- *Justify the request.* Units must articulate why the request is important. For greater priority, units may try to tie the requests for information to a PIR.
- *State accurately the latest time the information will be of value.* Units should state when information will no longer be useful, being truthful about the date. The information that units provide affects collection management and assets dedicated for higher-priority missions.

INTELLIGENCE PREPARATION OF THE OPERATIONAL ENVIRONMENT

5-5. The basis of intelligence support to IO is the IPOE process, a prerequisite to planning any operation. The mechanics of analyzing the information environment and adversary operations in the information environment are generally the same as those established to support IPOE for other military planning. Ideally, the J-2/G-2/S-2 has the lead on conducting IPOE and will include IO considerations in the analysis. However, IO planners can expect to assist in the process or conduct portions of the IPOE that are specific to IO. In such a case, J-2/G-2/S-2 products should be used as the basis for any IO-oriented analysis.

5-6. Information IPOE differs from traditional IPOE in purpose, focus, and end state. The purpose of IPOE is to gain an understanding of the information environment in a specific geographic area and to determine how the adversary will operate in that environment. The focus is on analyzing the adversary's use of information to gain an advantage. The end state is the identification of adversary vulnerabilities that friendly forces can exploit with IO, and adversary capabilities in the information environment against which friendly forces must defend.

5-7. For IO, IPOE results in the production of a graphic visualization product known as the CIO. The CIO is a map of the information environment that shows where and how information content and flow will affect military operations.

VISUALIZING THE INFORMATION ENVIRONMENT

5-8. To employ IO properly, commanders and staffs must grasp the character and impact of the information environment in their operational area. To do this, it is necessary to rationally analyze the information environment using the IPOE process and the three-dimensional model of the information environment (as described in Chapter 1).

5-9. Every operational area has an information environment with information moving through it. This information flow (the information domain) creates tangible, real-world effects by converting real-world situations (the physical dimension) into human perceptions that form the basis of individual and organizational behavior (the cognitive dimension).

5-10. Visualization of the information environment begins with the identification of its significant characteristics. To do this, planners must examine the operating area and identify the existing and projected characteristics that are relevant to the content and flow of information (for example, Step 1 of IPOE).

Chapter 5

Although there is no single set of characteristics useful for analyzing every information environment, some broad characteristics that can serve as starting points are terrain (weather), populace, civilian information infrastructure, civilian population, and third-party organizations. Figure 5-1 provides a broad list of characteristics of the information environment.

Significant Characteristic	Elements of the Characteristic	Information Requirements
Terrain	Those aspects of terrain and geography that impact information content and flow.	• How does terrain (and weather conditions) canalize and compartmentalize information content and flow? • How does terrain (and weather conditions) impact information flow?
Civilian Information Infrastructure	Key information system links and nodes (information conducts) in the operating area.	• What are the key information systems (telephone, microwave, Internet, and so on)? • What information content is passed on each information system? • Who (friendly forces, enemy, civilians, other organizations) uses each information system? • Who manages and controls the information systems?
Media	Radio, TV, print, and Internet, to include audiences.	• What media sources are available (for adversary and friendly use) in the AO? • What information content is reported by each media source? • Who is each media's audience? • What is the context or bias of the media outlets?
Civilian Population	• Demographics, such as distribution, language, religion, ethnicity, and education. • Cultural factors, such as societal structures, ideologies, perceptions, and beliefs.	• How does the populace communicate? • What information content does the populace need or want? • What are the populace's biases? • What is the populace's social organization? • What are the populace's cultural characteristics?
Third-Party Organizations	Interagency, nongovernmental organizations, private volunteer organizations, and international organizations that can be competing influences in the information environment.	• Who are the interagency, nongovernmental organizations, private volunteer organizations, and international organizations in the AO? • What are their purpose, goals, and objectives? • What information do these organizations project?

Figure 5-1. Visualization of the information environment

5-11. Planners analyze each of the previously identified significant characteristics using the three-dimensional model to determine specific impacts on operations in the information environment (that is, Step 2 of IPOE). To accomplish this task, each characteristic is considered within the framework of the three following dimensions:
- *Physical.* Units focus on what information systems in the operational area collect, process, and disseminate information. Identification should include the tangible aspects of each significant characteristic, such as technical information systems and networks (for example, radio towers, fiber-optic networks, and telephone networks) and nontechnical (human) information network nodes and links (such as persons with influence, key leaders, and face-to-face communications networks). Additionally, analysis should show where those information systems and networks are located in the physical environment.
- *Cognitive.* Units focus on the values, beliefs, and perceptions of key individuals and organizations in the operational area that make decisions, as well as how those decisions are formulated. This analysis should show how this human mental programming affects the value of specific information to those key individuals and organizations in the operational environment.
- *Information.* Units focus on how information flows and the content of that information. Flow describes the exchange of information in terms of conduits, form, and speed. Content includes the major subjects or topics circulating in the AO.

THE COMBINED INFORMATION OVERLAY

5-12. Analysis of the information environment should result in a CIO. The CIO (Figure 5-2, page 5-6) is not a static document; it is intended to be a working product that is continually refined as new information becomes available. Building a CIO begins with a map of the operational area (ideally the same map used by the intelligence and operations staffs). The information environment's significant characteristics are combined and plotted on the map to show aggregate effects in relation to the terrain of the operational area. This should result in the identification of subinformation environments.

5-13. Subenvironments are areas in which the information environment's significant characteristics and effects notably differ from adjacent areas. Because the composition of the information environment is not uniform, there will be distinct subinformation environments in the operating area. Physical features and cognitive aspects of the information environment determine subinformation environments. Leaders must consider that subinformation environments may transect international borders and unit boundaries. For example, subinformation environments may be based on the significant characteristics of ethnicity, media presence, and population density. One subenvironment may have a single ethnic group with widespread access to media and information, whereas another subinformation environment may have an entirely different populace group with limited or no access to outside media. The subenvironments can be further analyzed to determine their composition and character. Ideally, analysis will identify those parts of the operational area that favor either the friendly or adversary forces' operations.

5-14. After subenvironments are identified, key information nodes are selected within each subinformation environment. Information nodes are places, persons, or infrastructure that shape information content and flow by creating or transmitting information. Identifying key nodes is important because these areas may be the most-effective means for inserting messages into the local populace or adversary networks. The nodes critically affect information content and flow. Each subinformation environment will likely have one or more key information nodes. An information node can be human (for example, key communicators or leaders), technological (for example, cellular telephone towers, media outlets, and religious or meeting centers), or both. Nodes critical to both are key terrain in that they critically affect information within the operational environment and provide an advantage to either adversary or friendly forces. For example, a well-known mosque with an influential imam is a possible candidate for a key node because it creates or perpetuates information content affecting military operations.

Chapter 5

5-15. The CIO is a guide, not a rigid template. It typically depicts mission-significant aspects of the information environment, subinformation environments, key information nodes, and information flow in the operating area. The information included in the graphic should be presented in a concise manner. Whatever final form the CIO takes, it must present an operationally-relevant overview of the information environment. Every CIO will be unique because every information environment is different.

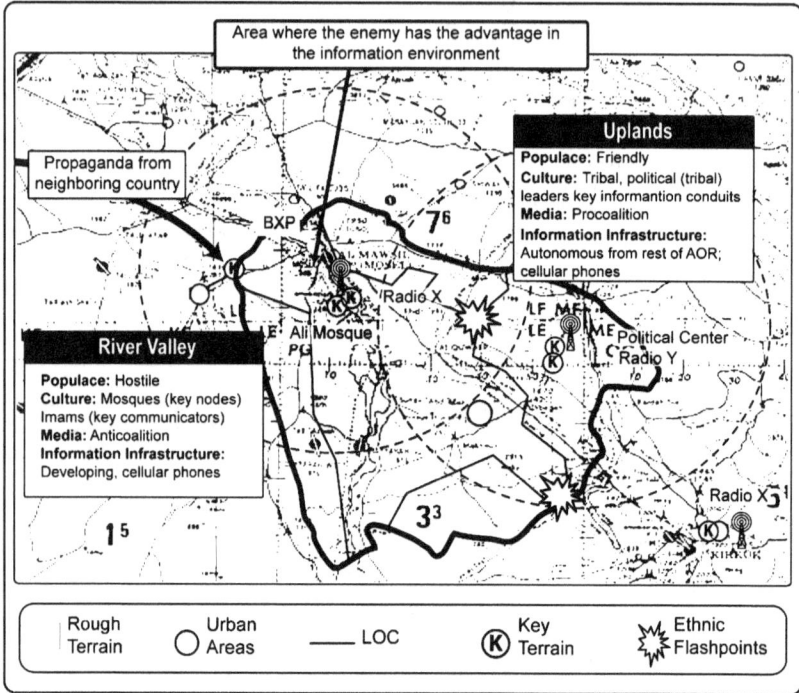

Figure 5-2. Example combined information overlay

ADVERSARY OPERATIONS IN THE INFORMATION ENVIRONMENT

5-16. Information and the information environment are not benign and often favor one side over another. Opposing forces use the information environment just as they use the physical environments of air, land, and sea to place their enemy at a disadvantage and to achieve their objectives. Understanding this, IO planners must identify how the adversary views and uses the information environment.

5-17. The adversary does not use the information environment in the same way or have the same constraints and means as U.S. forces. To avoid mirror-imaging the friendly concept of IO upon the adversary and to prevent mismatching U.S. capabilities and vulnerabilities, adversary operations in the information environment can be viewed in terms of activities to collect, protect, and project information. These three functions are universal to any armed force's ability to use information as combat power regardless of its organization, capabilities, and mission. As such, they form the basis of the adversary's capabilities (and vulnerabilities) in the information environment. Collect, protect, and project are defined as:

- *Collect.* To plan and execute operations, the adversary must collect accurate and timely information.
- *Protect.* To be successful, the adversary must protect its critical information from collection and maintain its means of communication.
- *Project.* To further its goals and objectives, the adversary must project the information into the information environment to influence the perceptions of its target audiences.

5-18. Depending on the adversary, the means used can be as simple as direct human observation and open sources (collect); couriers and intimidation (protect); and night letters, other printed materials, and graffiti (project). Ideally, analysis of how the adversary operates in the information environment is based on modeling, or templating.

TEMPLATING USING CENTER-OF-GRAVITY ANALYSIS

5-19. Once a clear understanding of the environment is established, IO planners should analyze adversary capabilities, requirements, and vulnerabilities in the information environment (Figure 5-3). The purpose of performing a COG analysis is to determine and evaluate the adversary's critical vulnerabilities for exploitation. Because this tool is used to evaluate the adversary, the appropriate time to perform this analysis is during Step 3 (evaluate the threat) of IPOE. The results of the COG analysis are later used during COA development to exploit identified vulnerabilities.

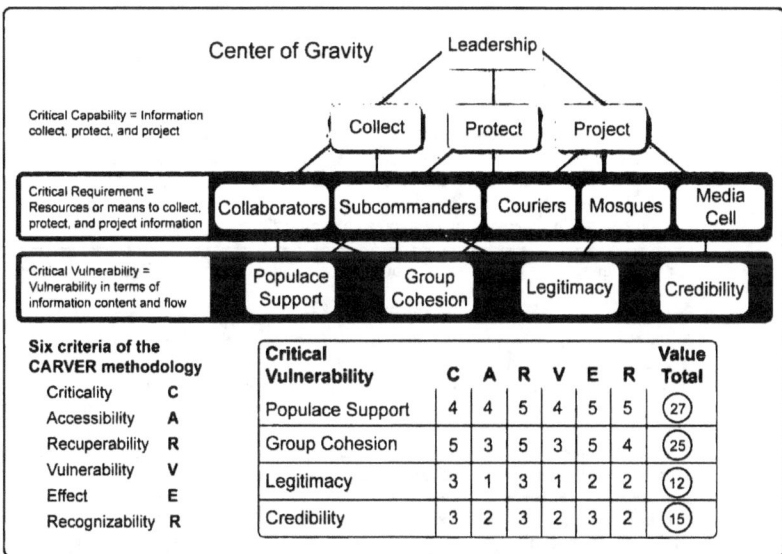

Figure 5-3. Example center-of-gravity analysis and the use of the CARVER process to rank and plot critical vulnerabilities in the information environment

5-20. The COG analysis of the adversary should be conducted by the J-2/G-2/S-2. If not, IO planners and operational detachments in the field can use a COG analysis to analyze the adversary in the information environment by—

- *Identifying potential threat COGs.* Visualize the threat as a system of functional components. Based upon how the threat organizes, fights, makes decisions, and its physical and psychological strengths and weaknesses, select the threat's primary source of moral or physical strength, power, and resistance. Depending on the level (strategic, operational, and tactical), COGs may be tangible entities or intangible concepts. To test the validity of the COG, the question that needs to be asked is: "Will the destruction, neutralization, or substantial weakening of the COG result in changing the threat's COA or denying its objectives?" When possible, the J-2/G-2/S-2 identifies the COG. If these assets are unavailable, then an independent information environment may need to identify the COG. Typically this is the adversary's information position, which is a way of describing the quality of information an organization possesses and its ability to use that information.
- *Identifying critical capabilities.* Each COG is analyzed to determine what primary abilities (functions) the threat possesses in the context of the operational area and friendly mission that can prevent friendly forces from accomplishing the mission. Critical capabilities are not tangible objects; rather, they are threat functions. To test the validity of a critical capability, the questions that need to be asked are: "Is the identified critical capability a primary ability in context with the given missions of both threat and friendly forces? Is the identified critical capability directly related to the COG?" A critical capability is a means that is a crucial enabler for a COG to function and, as such, is essential to the accomplishment of the adversary's specified or assumed objectives.

Note. The adversary's critical capabilities are the functions in the information environment—collect, protect, and project.

- *Identifying critical requirements for each critical capability.* Each critical capability is analyzed to determine what conditions, resources, or means enable threat functions or mission. To test validity of a critical requirement, the questions that need to be asked are: "Will exploitation of the critical vulnerability disable the associated critical requirement? Does the friendly force have the resources to affect the identified critical vulnerability?"

Note. Critical requirements usually are tangible elements such as communications means, nodes, or key communicators.

- *Identifying critical vulnerabilities for each critical requirement.* Each critical capability is analyzed to determine which critical requirements (or components thereof) are vulnerable to neutralization, interdiction, or attack. As the hierarchy of critical requirements and critical vulnerabilities are developed, interrelationships and overlapping between the factors are sought to identify critical requirements and critical vulnerabilities that support more than one critical capability. When selecting critical vulnerabilities, a critical-vulnerability analysis is conducted to pair critical vulnerabilities against friendly capabilities.

Note. Critical vulnerabilities may be tangible structures or equipment, or intangible perception, populace belief, or susceptibility.

- *Prioritizing critical vulnerabilities.* The CARVER is a special operations forces methodology to prioritize targets. The methodology can be used to rank-order critical vulnerabilities, thereby prioritizing the targeting process. The six criteria are applied against the critical vulnerability to determine impact on the threat organization as follows:
    - *Criticality* is the estimate of the critical vulnerability importance to the enemy. Vulnerability will significantly influence the enemy's ability to conduct or support operations.

- *Accessibility* is the determination of whether the critical vulnerability is accessible to the friendly force in time and place.
- *Recuperability* is the evaluation of how much effort, time, and resources the enemy must expend if the critical vulnerability is successfully affected.
- *Vulnerability* is the determination of whether the friendly force has the means or capability to affect the critical vulnerability.
- *Effect* is the determination of the extent of the effect achieved if the critical vulnerability is successfully exploited.
- *Recognizability* is the determination if the critical vulnerability, once selected for exploitation, can be identified during the operation by the friendly force, and can be assessed for the impact of the exploitation.

5-21. The result of the analysis should determine the adversary's vulnerabilities that can be attacked by friendly-force IO capabilities. Figure 5-3, page 5-7, provides a visual depiction of the relationship between critical vulnerabilities.

5-22. In testing the validity of the COG analysis, leaders should apply the following questions:
- Will destruction, neutralization, or substantial weakening of the COG result in changing the threat's COA or denying its objective?
- Does the friendly force have the resources and capabilities to accomplish destruction or neutralization of the threat COG? If the answer is no, than the threat's identified critical factors must be reviewed for other critical vulnerabilities, or planners must reassess how to attack the previously identified critical vulnerabilities with additional resources.

ADVERSARY ACTIVITIES IN THE INFORMATION ENVIRONMENT

5-23. As part of determining the adversary's COA (Step 4 of IPOE), IO planners should determine how the threat employs its assets to operate in the information environment and achieve information superiority over U.S. forces. To be valid, this analysis should be developed in concert with, and integrated into, the intelligence staff's analysis.

5-24. To graphically depict an adversary's COA in the information environment, planners start with the CIO and then add why (likely information objectives and actions), where (location of primary information assets and means), when (a forecast of when the adversary will employ its assets), and how (the employment of capabilities) the adversary will seek information superiority. The result is a concept of the operation that describes how the adversary will operate in the information environment. In turn, this product can be used during mission planning.

CONSIDERATIONS

5-25. Intelligence that may be considered less-than-credible or insignificant to a traditional intelligence analyst can be key to an IO planner. Some examples include the following:
- *Perceptions*. Planners use the target audience's existing perceptions to their advantage. Knowledge and understanding of existing perceptions of the population, insurgent groups, and HN government and forces can provide IO opportunities to exploit to achieve desired effects or counter the adversary's exploitation. Gaps in understanding perceptions can be answered using MIS teams, CA teams, and during the conduct of key-leader engagements. Examples of perceptions to exploit or counter include—
  - An insurgent group that believes a mole exists within their organization (exploit).
  - The populace believes that insurgents are forcing U.S. forces out of bases (counter).
- *Rumors*. Planners use rumors as a method to achieve an effect. IO planners look for various fissures in organizations to exploit and shape perceptions. Examples are—
  - Mistrust or jealousy among individuals.
  - Greed or desire for power.

- *Sensitive-site exploitation.* Soldiers conducting sensitive-site exploitation, should observe the location and mentally record information that may not be of immediate tactical value but can be used by IO planners to better understand the information environment. Questions that Soldiers should ask include—
    - Is there a TV? What channel is it on?
    - Is there a radio? What station is it on?
    - Are there periodicals? If so, what type are they? What language?
    - Are there music compact discs? What type of music?
    - How much food is in the house? Is it more than necessary to support the family?

Appendix A
Planning Aids

This appendix provides multiple planning aids for Soldiers conducting IO missions. It outlines the duties and responsibilities of the IOWG and provides Soldiers with numerous mission analysis tools.

INFORMATION OPERATIONS WORKING GROUP

A-1. The IOWG brings together representatives of those staff elements concerned with the information operation. It is the most important meeting held by the assistant chief of staff, information operations (G-7)/information operations staff officer (S-7). The unit SOP should address the following for the working group:
- *Purpose.* The purpose of the IOWG is to synchronize the contributions of all staff elements to the work of the IO section.
- *Frequency.* The frequency of IOWGs depends on the situation and echelon. The working group may gather daily, weekly, or monthly, depending on the situation, echelon, and time available. Corps and division HQ may have daily (combat operations) or weekly (stability operations) IOWGs. Battalion and brigade HQ normally have fewer working groups than higher echelons.
- *Composition (chair and attendees).* The G-7/S-7 determines participation in the IOWG. It is a mix of staff-element representatives and subject-matter experts.
- *Inputs and outputs.* Attendees must know what information, products, and formats they are required to produce and use.
- *Agenda.* The formality of the IOWG also varies by echelon. For purposes of organization and focus, even the simplest IOWG should have an agenda.

COMPOSITION

A-2. The composition of the IOWG is tailored to the agenda. Representatives from every staff section need not attend each and every IOWG. Participants are selected because they either represent a critical element or capability, or have expertise that is critical to the IO. Core participants are staff members and subject-matter experts who regularly attend the IOWG due to their role in IO. Core participants include the following:
- IO personnel.
- MIS representatives.
- EW representatives.
- OPSEC representatives.
- COCAM representatives.
- CA representatives.
- PA representatives.
- MILDEC representatives.
- Representatives from the J-2/G-2/S-2.
- Representatives from the J-3/G-3/S-3, effects cell.
- Special technical operation planners.
- Fire support officer.
- Others, as required.

Appendix A

A-3. There are other staff members who may not attend the IOWG on a regular basis, but whose role is no less important. They include the following:
- COMCAM officer in charge.
- G-6/S-6 representative.
- Cultural advisor.
- Chaplain.
- Political advisor.
- Subordinate-unit IO officers.
- Staff Judge Advocate representative.
- Liaison officers.

DUTIES AND RESPONSIBILITIES

A-4. Figure A-1 provides the duties and responsibilities of the IO working group members.

G-7/S-7	• Chair and facilitate working group. • Establish and enforce agenda. • Encourage active participation.
IO Capability Representatives	• Serve as subject-matter expert for their staff function or unit. • Provide input on capability status. • Provide input on current and future tasks and activities.
G-2/S-2 Representatives	• Provide intelligence relevant to IO. • Answer working group requests for information.
G-3/S-3 Representatives	• Provide input on current and future operations.
Subordinate Unit IO Representatives/Liaison Officers	• Serve as subject-matter expert for their unit. • Provide input on current and future missions, priorities, and tasks.
Recorder	• Record, write, and disseminate minutes of working group.
Other Participants	• Serve as subject-matter expert for their staff function or area of expertise. • Actively participate in the working group.

Figure A-1. Information operations working group duties and responsibilities

PREPARATION

A-5. Preparation is critical to a successful IOWG. A successful working group requires a collective effort from the IO section. For example, someone sets and prepares the agenda, another person notifies participants and ensures each is prepared to provide meaningful input to the working group, and another person prepares the IOWG presentation. Preparation tasks for the IWOG include the following:
- Set agenda.
- Notify participants:
    - Verify time and place of IOWG.
    - Identify additional participants.
- Review status of due-outs and contact those participants with due-outs.
- Coordinate with participants who have formal input.
- Publish a read-ahead packet:
    - If possible, provide IOWG materials to participants prior to the meeting.
    - Ensure participants provide input to IOWG presentation prior to the meeting.
- Assign a recorder to take minutes for the working group.

A-6. There are certain basics of meeting management that—if applied to the IOWG—can increase its effectiveness. Some basic suggestions include the following:
- Meet at established times and places.
- Keep meetings short—1 hour is a good rule of thumb.
- Have an agenda and follow it.
- Tailor working group membership to those people who are truly needed.
- Encourage participation by members; working groups are not one-way conversations.
- Complete detailed work and coordinate actions before the IOWG. Discuss actions and issues that are relevant to the working group.
- Identify and work critical issues. Identify and work side issues after the working group.
- Follow through on actions and due-outs. Record and track the results of the working group and publish minutes.
- Insist on timely delivery of due-outs and products.
- Invite subordinate and higher-command representatives.
- Give feedback to working-group members.

AGENDA

A-7. IOWG agendas vary by mission, situation, and echelon. A typical IOWG agenda includes the following:
- Roll call.
- Due-outs from previous IOWG.
- Intelligence update.
- Assessment update.
- Operations update.
- Discussion and issues.
- Review of due-outs.
- Conclusion.

A-8. Some IOWGs are organized along the lines of a targeting meeting, whereas others are similar to an operations meeting. Regardless of what agenda the IOWG takes, the purpose remains the same—to synchronize IO's contributing capabilities.

Due-Outs from Previous Working Group

A-9. Due-outs address unanswered questions or issues from the previous IOWG. Previous due-outs not answered during the IOWG should be carried over to the next IOWG for resolution. Typically, a due-out identifies the issue or question requiring resolution, and the person or element responsible for answering the due-out.

Intelligence Update

A-10. The purpose of the intelligence update is to answer current G-7 intelligence requirements. As such, it focuses on the information environment, the adversary's actions in the information environment, and the impact of those actions on friendly operations. Intelligence updates for IO should not be a regurgitation of other conventional intelligence updates. One way to structure the intelligence update is to capture significant events in the information environment and organize them by the G-7's intelligence requirements. Figure A-2, page A-4, provides a sample intelligence update format.

Assessment Update

A-11. The purpose of the assessment update is to assess the impact and effectiveness of current IO. Its focus is on analyzing and presenting information and intelligence from unit operations and intelligence reports, as well as input from the IOWG members. Figure A-3, page A-4, provides a sample of how the

Appendix A

assessment update can be depicted. Each operational area has a pie chart that represents the status of the current IO objectives (in this example there are five IO objectives).

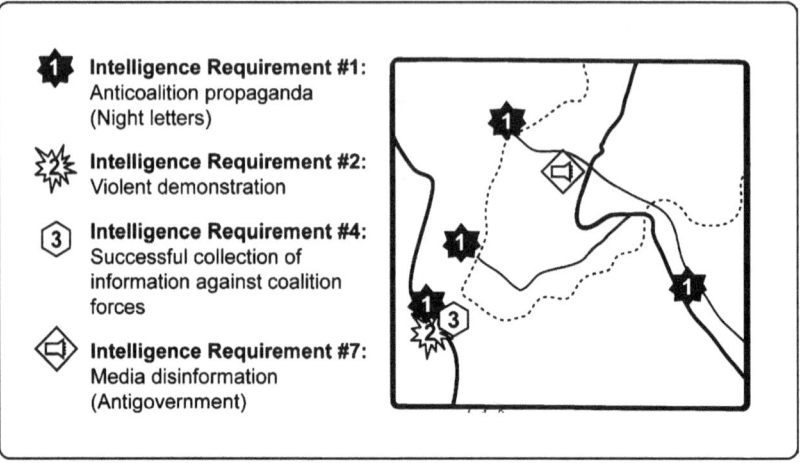

Figure A-2. Sample intelligence update format

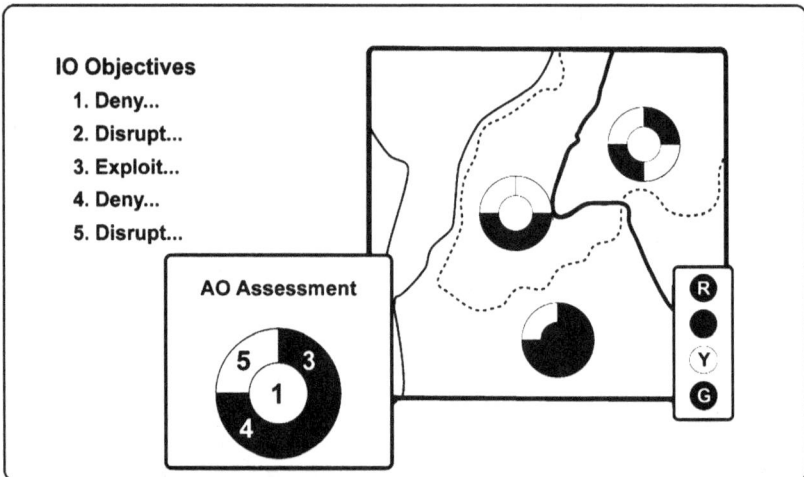

Figure A-3. Sample assessment update format

OPERATIONS UPDATE

A-12. The purpose of the operations update is to synchronize the IO objectives with element/capability tasks and targets for current and future (mid-range) IO. The focus is on gaining or maintaining information superiority. One way to structure the operations update is to use graphics that show time, location, and purpose for key IO tasks for each major operation. Figure A-4 provides a sample format.

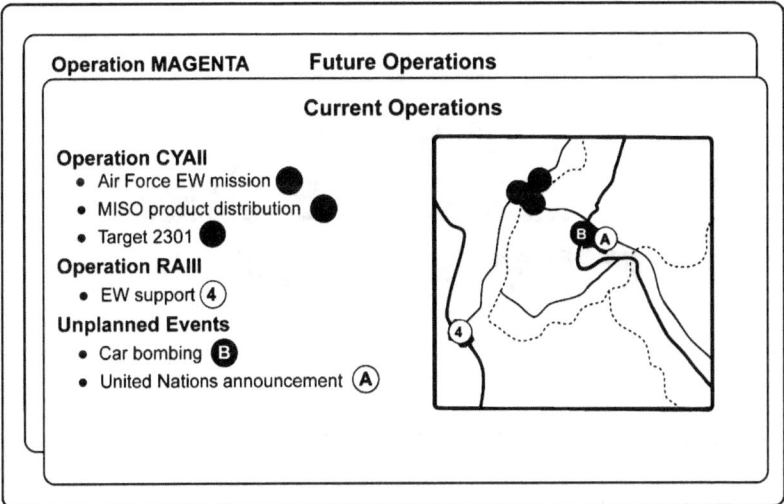

Figure A-4. Sample operations update format

Discussion and Issues

A-13. The purpose of the discussing issues or special topics is to support the G-7/S-7 decisionmaking and to synchronize the current and future activities of IO supporting capabilities. Discussion topics are selected by the G-7/S-7. Working-group participants have the opportunity (and responsibility) to discuss the topics from the perspective of their staff function or area of expertise. This discussion can be facilitated or focused by the use of an operations calendar (Figure A-5, page A-6) containing critical events and planned operations.

Review of Due-Outs

A-14. The purpose of reviewing due-outs is to ensure the working group participants understand and acknowledge their due-outs and responsibilities for the next meeting. Prior to final questions and comments, the G-7/S-7 reviews new due-outs identified during the working group as well as any open due-outs from the previous working groups. Each due-out should identify the issue or question requiring resolution, and the person or element responsible for answering the due-out.

Conclusion

A-15. The G-7/S-7 briefly discusses what the meeting accomplished and what working-group objectives were met. If necessary, side conversations, meetings, and other subworking groups are identified and scheduled.

Appendix A

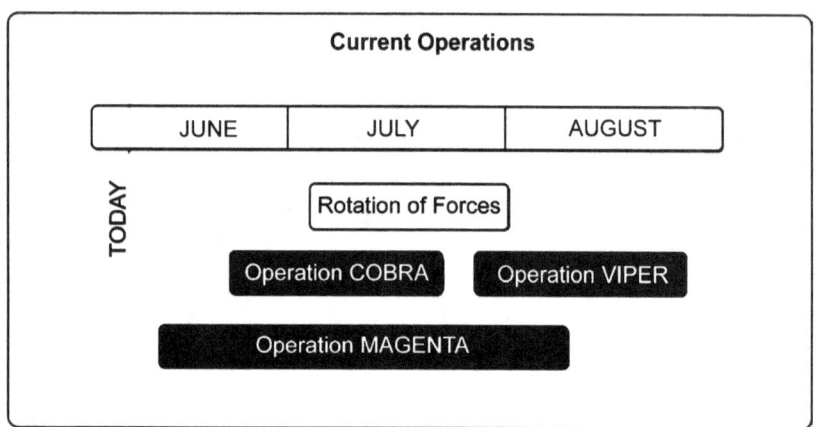

Figure A-5. Sample operations calendar

INFORMATION OPERATIONS PLANNING AIDS

A-16. Figure A-6 is an IO planning aid depicting the relationship between the military decisionmaking process and IO. Figure A-7, page A-7, depicts the relationship between IPOE and IO.

Military Decisionmaking Process Step	Information Operations Focus
Receipt of Mission	• Conduct initial assessment of information operation. • Determine IO planning requirements.
Mission Analysis	• Understand IO situation. • Analyze the higher HQ IO. • Define and analyze the information environment and threat. • Develop IO mission statement and objectives. • Seek commander's IO guidance.
COA Development	• Identify friendly IO capabilities and vulnerabilities. • Develop IO concept of support.
COA Analysis	• Visualize operations in the environment. • War-game IO concept of support against how the enemy will employ its information systems and assets.
COA Comparison	• Analyze and evaluate IO support to each COA.
COA Approval	• Finalize details of the information operation.
Orders Production	• Prepare IO annex and input to base operation order/operation plan.

Figure A-6. Information operations planning aid

A-17. The purpose of performing a COG analysis (Figure A-8, page A-7) is to determine and evaluate the adversary's critical vulnerabilities for exploitation. Because this tool is used to evaluate the adversary, the appropriate time to perform this analysis is during Step 3 (evaluate the threat) of IPOE.

Planning Aids

A-18. In testing the validity of the COG analysis, leaders should apply the following questions:
- Will destruction, neutralization, or substantial weakening of the COG result in changing the threat's COA or denying its objective?
- Does the friendly force have the resources and capability to accomplish destruction or neutralization of the threat COG? If the answer is no, then the threat's identified critical factors must be reviewed for other critical vulnerabilities, or planners must reassess how to attack the previously identified critical vulnerabilities with additional resources.

Intelligence Preparation of the Operational Environment Steps	Information Operations Focus	Analysis Product
Define the operational environment.	Define the information environment.	Combined information overlay—significant characteristics of the information environment and effects on operations.
Describe the operational environment's effects.	Describe the information environment's effects.	
Evaluate the threat.	Evaluate the threats' information system.	• Threat COG analysis—critical vulnerabilities. • Threat templates—who makes decisions; what nodes, links, and systems the threat uses; how information assets are employed.
Determine threat COAs.	Determine threat actions in the information environment.	Information situation template—when, where, and why the threat will seek to gain information superiority.

Figure A-7. Relationship between intelligence preparation of the operational environment and information operations

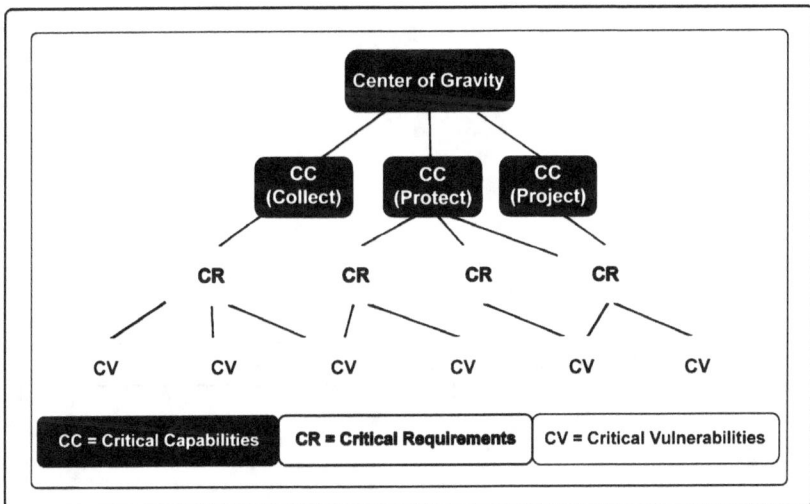

Figure A-8. Sample center-of-gravity analysis

A-19. Figure A-9, page A-9, is a sample combined information overlay. Figure A-10, page A-9, depicts a sample IO mission-to-task product chart. Figure A-11, page A-10, depicts a sample IO mission and tasks (tactical level). Figure A-12, page A-10, depicts a sample COA sketch.

A-20. In addition to doctrinal effects, IO have a number of nondoctrinal effects, to include the following:
- *Destroy.* This renders a target so damaged that it cannot function as intended nor be restored to a usable condition without being rebuilt.
- *Degrade.* This reduces the effectiveness or efficiency of adversary information systems, assets, or functions.
- *Disrupt.* This temporarily interrupts the flow of information.
- *Deceive.* This misleads or manipulates adversary understanding of friendly forces' activities, capabilities, vulnerabilities, and intentions.
- *Influence.* This affects an adversary or others perceptions, attitudes, and behavior to support friendly-force objectives.
- *Preserve (nondoctrinal).* This maintains the effectiveness or efficiency of friendly-force information systems, assets, or functions (related to doctrinal effect *protect*).
- *Deny.* This hinders or prevents an adversary and others from gaining access to, collecting, or using information concerning friendly forces.

A-21. Possible IO tasks include the following:
- Control.
- *Counter.*
- Counter-reconnaissance.
- Defeat.
- Delay.
- *Demonstrate.*
- Destroy.
- *Deter.*
- *Engage.*
- Fix.
- *Inform.*
- Interdict.
- Isolate.
- *Jam.*
- Neutralize.
- *Persuade.*
- *Prevent.*
- Protect.
- Secure.
- Suppress.

Note. Italicized tasks are proposed IO tactical tasks.

Planning Aids

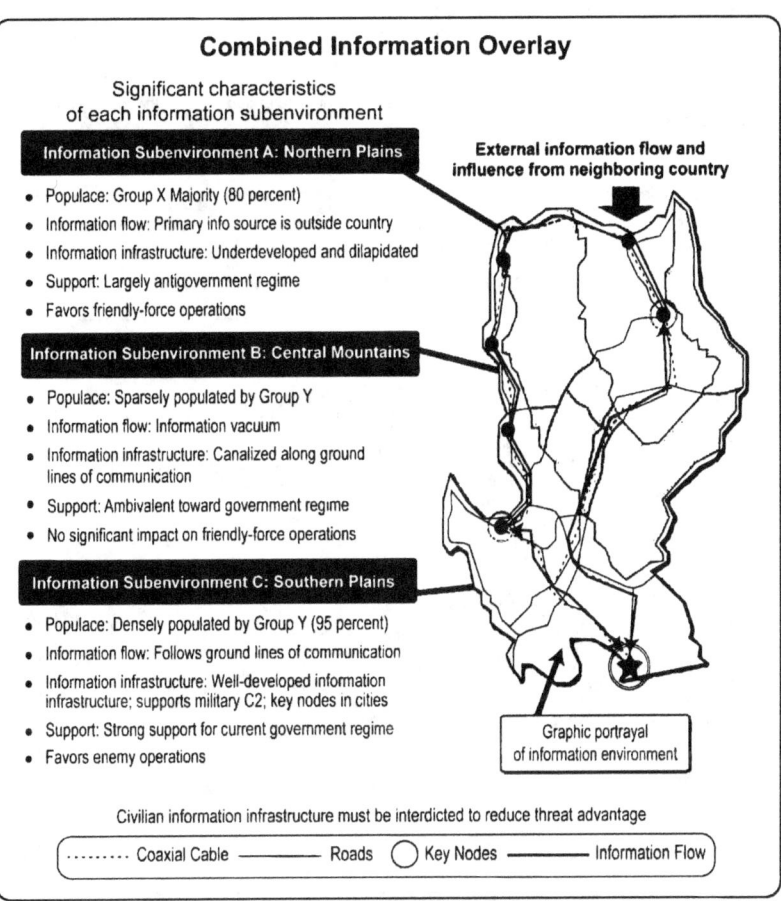

Figure A-9. Sample combined information overlay

Mission	Tasks Products
IO Mission Statement	How IO will support the command's mission (who, what, where, when, why).
IO Objectives (3 to 5 Objectives per Phase)	What IO will do to affect the information environment (effect, object of the effect [target], purpose of the effect).
Tasks to IO Elements	What actions the elements will perform to execute the information operations (task, purpose).
Tasks to Units	Task, purpose.
IO Concept of Support	How the information operation will be conducted (commander's intent for IO, information superiority for the operation, general plan for IO, priority).

Figure A-10. Mission-to-task products

Appendix A

IO Mission: On order, the friendly force disrupts the enemy ground and air defense forces' C2, influences civilian populace perceptions, and protects Corps' critical information in the AO to facilitate destruction of 1st Operational Strategic Command forces.

IO Objectives:

- Disrupt the enemy force's air defense C2 to prevent coordinated engagement of the friendly force's deep attacks.
- Disrupt operational reserve command posts and communication networks to delay employment of reinforcing or counterattack forces.
- Influence civilian populace in occupied areas to minimize interference with the friendly force's operations.
- Deny detection and identification of the friendly force's main and tactical command posts to prevent targeting by the enemy force's artillery fires.

Figure A-11. Sample information operations mission and tasks (tactical level)

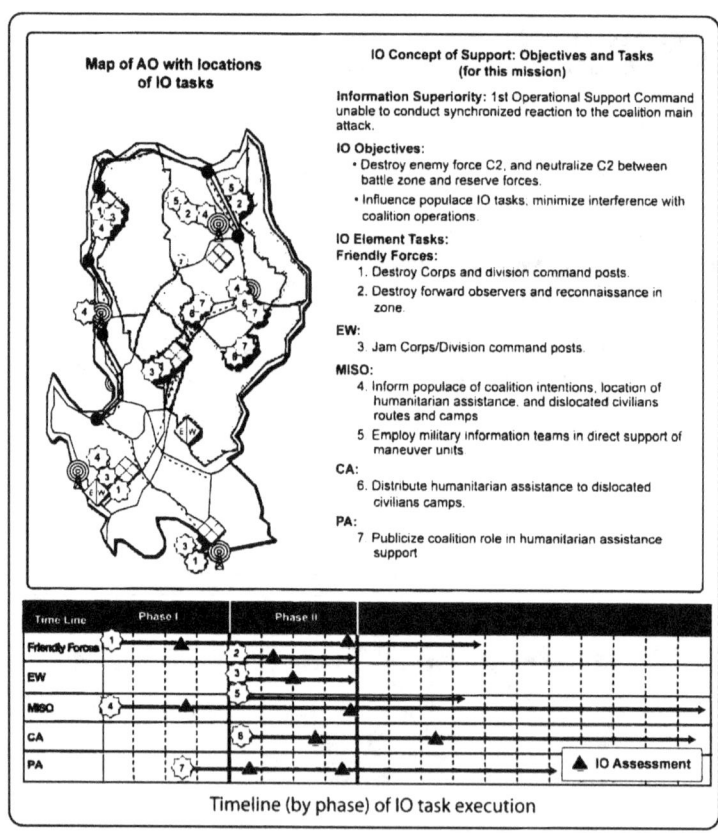

Figure A-12. Sample course-of-action sketch

MISSION ANALYSIS AND INFORMATION OPERATIONS

A-22. As part of the planning process, the IO staff must conduct its own mission analysis. Figure A-13 depicts the relationship between IO and mission analysis.

Military Decisionmaking Process Step	Information Operations Focus
Analyze Higher HQ Order	Analyze the higher HQ IO.
Perform IPOE	Define information environment and determine threat COAs.
1. Determine Tasks	Determine what the IO must do.
2. Review Available Assets	Determine organic and support IO capabilities.
3. Determine Constraints	Determine constraints on information content and flow.
4. Identify Facts and Assumptions	Identify facts and assumptions relevant to information content, flow, and use.
5. Perform Risk Assessment	Input hazards resulting from IO tasks.
6. Determine CCIR and EEFI	Determine EEFI.
Determine Intelligence, Surveillance, and Reconnaissance Plan	Input information requirements for IO.
Update Timeline	Input lead time for IO tasks.
Write Restated Mission	Write IO mission statement (if used).
7. Deliver Mission-Analysis Briefing	Input to mission-analysis briefing.
Approve Restated Mission	Approve IO mission statement (if used).
Develop Commander's Intent	Input to commander's intent.
Issue Commander's Guidance	Issue guidance for IO.
Issue Warning Order	Input for IO.
Review Facts and Assumptions	Address changes to IO planning factors.

Figure A-13. Mission analysis and information operations

A-23. Figure A-14, page A-12, provides a sample mission-analysis work sheet. The mission-analysis work sheet—
- Provides a tool to conduct mission analysis.
- Focuses on the minimum information needed for a plan.
- Follows the sequence of the mission-analysis briefing format, not the steps of mission analysis.

A-24. Identify specified and implied tasks to IO; not tasks to the capabilities. Tasks to the capabilities may be a constraint because they allocate resources away from the IO. Specified tasks are tasks specifically assigned to a unit by its higher HQ. Implied tasks are tasks that must be performed to accomplish a specified task or the mission, but are not stated in the higher HQ order.

A-25. IO tasks ignore staff coordination, administrative, and SOP tasks (for example, conducting a weekly IOWG, or submitting daily reports). Leaders organize identified specified and implied tasks to improve clarity. Tasks may be divided into two categories—
- Tasks to shape the information environment and to engage enemy forces.
- Tasks associated with information flow and information content.

A-26. Essential tasks are specified or implied tasks that must be executed to accomplish the mission. Leaders select three to five essential tasks, which are approved by the commander during the mission-analysis briefing.

Appendix A

1. **Facts:** Situation remains that the enemy will continue to use improvised explosive devices, small-unit attacks, and ambushes to attempt to gain strength for its cause. Friendly forces will continue to find and engage the enemy to eliminate improvised explosive devices. Current strength remains at 90 percent. Troops in the area of operations remain on the offensive and material readiness remains at 90 percent.
2. **Assumptions:** Attacks will continue to increase as the weather continues to improve, allowing for more insurgents and better leadership. This has been the pattern over the last several years and by all indications will continue.
3. **Tasks:** a. Specified: Detachments within the area of operations will continue to hunt for suspected insurgents and find, fix, capture, or eliminate as situation dictates. Additionally, detachments will continue to provide support to coalition forces and provide humanitarian support to the host nation. b. Implied: Continue offensive operations against insurgents. Continue to provide operational support to coalition forces and provide humanitarian support and security for the host nation. c. Essential: Provide for security and protect essential elements of friendly information. Provide training support to host-nation and coalition forces and conduct offensive operations to deter and eliminate insurgent activities in concert with collation partners.
4. **Constraints:** Close air support is authorized in rural areas. The use of close air support in urban areas must be approved by regional commander. Cultural support team will interrogate women and children only and detachments will provide security, as applicable.
5. **Available Assets:** All organic detachment equipment will be used, to include combat controllers, interpreters, Civil Affairs teams, Military Information Support teams, provincial reconstruction teams, and coalition support. Current rules of engagement remain in effect. Avoid using themes that will denigrate host-nation forces, such as cultural and religious themes. All releases to the media must first be approved by the public affairs office.
6. **Risk Assessment:** Improvised explosive devices will remain a constant threat. Weather conditions will hamper aviation and air support.
7. **CCIR:** In order to maintain the offensive, the commander needs to know current situational status and personnel status for both U.S. and coalition forces. a. PIR: Commander needs a clear and concise understanding of the enemy strengths and weaknesses; the disposition of forces and key leaders. b. FFIR: Commanders need a clear and concise understanding of coalition and friendly forces within the area of operations—specifically, which units are available to support detachments if needed, to include other ongoing operations within the detachment area of operations.
8. **EEFI:** Protection of intelligence requirements. Mission of higher and friendly forces within the area of operations, infiltration and exfiltration routes, unit strength, and communications must be protected.

Figure A-14. Example of a mission-analysis work sheet

REVIEW AVAILABLE ASSETS

A-27. Leaders examine available assets to determine IO capabilities and limitations, considering both organic and supporting assets based on current task-organization, support relationships, and status of units. Assets are compared to specified, implied, and essential tasks to determine if there are enough assets to accomplish all tasks. It may be useful to translate assets into IO capabilities (effects and targets). Figure A-15, page A-13, provides a sample matrix to help in this comparison.

Planning Aids

Organization	Asset	Means	Supported Essential Task	Effect	Target
Organic Assets					
Tactical Psychological Operations Company	1 x Product Development Detachment 9 x Tactical Psychological Operations Teams	Face-to-face Loudspeaker Handbills Posters Radio Television	N/A	Influence Inform Deceive	Local populace
Military Intelligence Company	3 x Prophet	Electronic attack	N/A	Disrupt	Enemy communication
Civil Affairs Battalion	12 x Tactical Support Teams	Face-to-face Humanitarian assistance Medical assistance Reconstruction projects	N/A	Influence Inform Co-opt	Local civilian leaders Local populace
Supporting Assets					
CFACC	EA6B EC130	Electronic attack Electronic support	Disrupt Degrade Destroy Inform Influence		Division-level communications Corps-level communications

Figure A-15. Example of an information operations asset/capability matrix

DETERMINE CONSTRAINTS

A-28. Constraints are restrictions on the use and employment of IO. The two types of constraints are prohibited actions (cannot do) and directed actions (must do; that is, resources and assets are required to do something). Constraints affect the use of IO capabilities and may be found in base orders, annexes, and appendixes. Constraints may be organized by affect on information content and flow (Figure 3-4, page 3-6).

FACTS AND ASSUMPTIONS

A-29. Facts and assumptions establish an understanding of the situation. Facts are known data concerning the situation. Assumptions are accepted as true in absence of facts. Leaders focus on facts and assumptions that concern assigned tasks. Figure A-16, page A-14 provides a sample fact and assumption analysis. Facts and assumptions may be organized as follows:
- Information environment (content and flow).
- Adversary capabilities and vulnerabilities in the information environment.
- Friendly capabilities and vulnerabilities in the information environment.

Information Environment	• Local populace is illiterate. (Fact) • Radio primary means to reach populace. (Fact) • Populace is pro-United States. (Fact) • Local leaders can control populace behavior. (Assumption)
Adversary Forces	• SIGINT is limited to short-range very high frequency radio. (Fact) • Use satellite and cell phones for C2. (Fact) • Will direct adversary information against U.S. forces. (Assumption)
Friendly Forces	• Friendly forces can jam the enemy's C2. (Fact) • Friendly forces can use local radio stations. (Assumption)

Figure A-16. Example of a fact and assumption analysis

RISK ASSESSMENT

A-30. Leaders identify and assess risks in the information environment arising from the essential tasks for IO. Risk assessment has five steps:
- Identify hazards (accomplished during mission analysis).
- Assess hazards (accomplished during mission analysis).
- Develop controls.
- Determine residual risk.
- Implement controls.

A-31. IO planners identify two kinds of hazards (risks). Tactical risk is concerned with hazards that exist because of the presence of the enemy or adversary. Accidental risk includes risks to friendly forces, to civilians, and the operation's impact on the environment.

INPUT TO COMMANDER'S CRITICAL INFORMATION REQUIREMENT

A-32. The CCIR identifies the information needed for direct execution of the mission. There are two types of CCIRs—PIRs and FFIRs:
- PIRs are information the commander must know about the enemy. For IO, PIRs focus on conditions in the information environment and threat actions to affect the information environment.
- FFIRs are information the commander must know about the friendly force. For IO, FFIRs focus on the friendly force's capability to shape information content and flow.

ESSENTIAL ELEMENTS OF INFORMATION

A-33. EEFI is information that must be protected from the adversary's intelligence system. Sources of information for developing EEFI are commander's guidance, facts, assumptions, and essential-task lists, and the intelligence estimate (information about adversary intelligence capabilities and requirements). EEFIs are written as statements, not questions; otherwise, EEFI may be confused with PIRs. Figure A-17, page A-15, provides an example.

PIR	<ul><li>What information systems is the adversary using for C2?</li><li>What means is the enemy using to disseminate propaganda?</li><li>Is adversary information turning popular opinion against operations?</li></ul>
FFIR	<ul><li>Media coverage of alleged friendly force's misconduct.</li><li>Civilian casualties caused by friendly-force operations.</li></ul>
EEFI	<ul><li>Friendly force's means of intelligence collection.</li><li>Tribal leaders who are assisting friendly forces.</li></ul>

Figure A-17. Example of a commander's critical information requirement and essential elements of information for information operations

MISSION-ANALYSIS BRIEFING

A-34. The IO portion of the briefing is included either in the G-3 and G-2 planners' presentations or, when appropriate, developed as separate slides. IO input typically includes the following:

- *Mission.* Commander's intent for IO of HQ two levels up and own commander's IO guidance.
- *IPOE.* CIO and enemy COAs in the information environment.
- *Facts and assumptions.* Critical facts and assumptions for IO.
- *Tasks.* Specified, implied, and essential tasks for IO.
- *Constraints.* Restrictions on the use and employment of IO.
- *Forces available.* Organic and supporting IO-capable assets and their capabilities and limitations.
- *Risk assessment.* Risks in the information environment.
- *CCIR.* Input to PIR, FFIR, and EEFI.
- *Timeline.* Input to the time allocation plan for accomplishment of IO essential tasks.
- *Restated mission.* IO mission statement (if used).

INFORMATION ENVIRONMENT CONSTRUCT

A-35. Figure A-18 depicts the different dimensions of the information environment and their key characters.

Information Environment Dimensions	Key Characteristics
Cognitive<ul><li>Individual and collective consciousness</li><li>Where decision are made</li></ul>	<ul><li>Beliefs</li><li>Values</li><li>Perceptions</li><li>Awareness</li><li>Decisionmaking</li></ul>
Information<ul><li>Intersection of physical and cognitive dimension</li><li>Where information is created and exists</li></ul>	<ul><li>Information content</li><li>Information flow</li><li>Information functions—collect, project, protect</li></ul>
Physical<ul><li>The physical world—land, sea, air, and space</li><li>Where information systems and networks reside</li></ul>	<ul><li>Technological information systems, Internet, media</li><li>Human—societal organization, military formations, third-party organizations</li></ul>

Figure A-18. Information environment

Appendix A

DEFINE INFORMATION ENVIRONMENT

A-36. Leaders define the information environment by examining the AO to identify the following significant characteristics of the cognitive, information and physical dimensions:
- *Terrain.* Canalization and compartmentalization.
- *Civilian information infrastructure.* Key links and nodes.
- *Media.* Radio, TV, print, and Internet, including audiences.
- *Civilian population:*
    - Demographics, such as distribution, language, religion, ethnicity, and education.
    - Cultural factors, such as societal structures, ideologies, perceptions, and beliefs.
- *Third-party organizations.* Nongovernmental organizations, private organizations, criminal organizations.

A-37. The information environment variances by level of war are depicted in Figure A-19.

Tactical	Physical	• Terrain and weather. • Local information systems. • Face-to-face contact.
	Information	• Line-of-sight flow. • Content addresses immediate needs.
	Cognitive	• Immediate perceptions and behavior.
Operational	Physical	• Regional information systems.
	Information	• Over-the-horizon flow. • Content addresses higher-level issues and concepts.
	Cognitive	• Near-term group perceptions and behavior.
Strategic	Physical	• Mass, long-distance information systems.
	Information	• Global flow. • Content addresses abstract ideas, ideologies, and philosophies.
	Cognitive	• Long-term perceptions and beliefs.

Figure A-19. Information environment variances by level of war

DESCRIBE THE INFORMATION ENVIRONMENT'S EFFECTS

A-38. Leaders analyze each significant information environment characteristic in detail and plot the data in a template (such as the sample provided in Figure A-20, page A-17).

Planning Aids

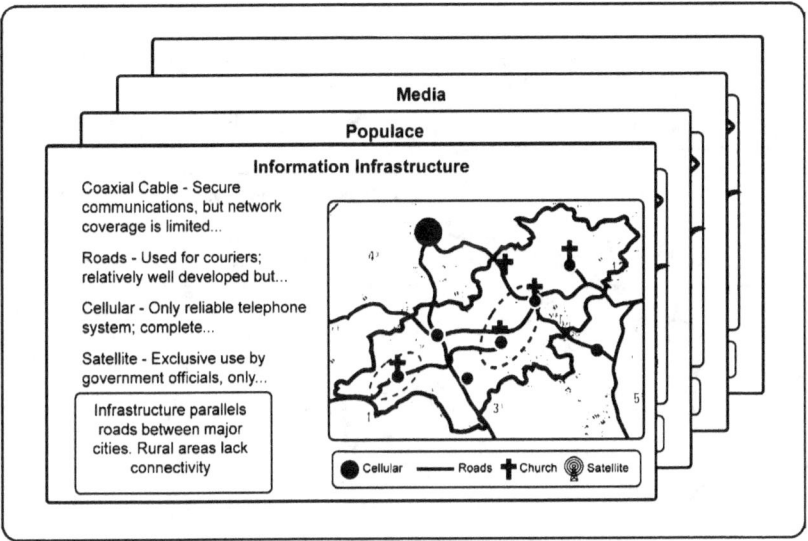

Figure A-20. Sample information environment effects matrix

DESCRIBE SUBINFORMATION ENVIRONMENT AND KEY NODES

A-39. By identifying and acting on key nodes, a military force can affect the information environment. Subinformation environments are areas in which the information environment's characteristics and effects are notably different from those of adjacent areas. Subinformation environments—
- Are determined by physical features and cognitive aspects of the information environment.
- Are formed by interactions of physical and cognitive dimensions.
- Determine an advantage to the friendly or adversary force.

A-40. Information nodes are key terrain in the information environment. Information nodes are places, persons, or infrastructures that shape information content and flow by creating or transmitting information. Information nodes—
- Exist in each subinformation environment.
- Can be human, technological, or both.
- Are located at the center of information content and flow.
- Critically affect information flow and content.
- Provide an advantage to one side or the other.

COMBINED INFORMATION OVERLAY

A-41. The CIO is a graphic depiction of where and how the information environment's effects will impact military operations. Figure A-21, page A-18, provides a template for a CIO. Figure A-22, page A-19, provides a sample CIO. The CIO—
- Depicts subinformation environments and key nodes.
- Describes information flow in the operating area.
- Includes a "so what" analysis.

Appendix A

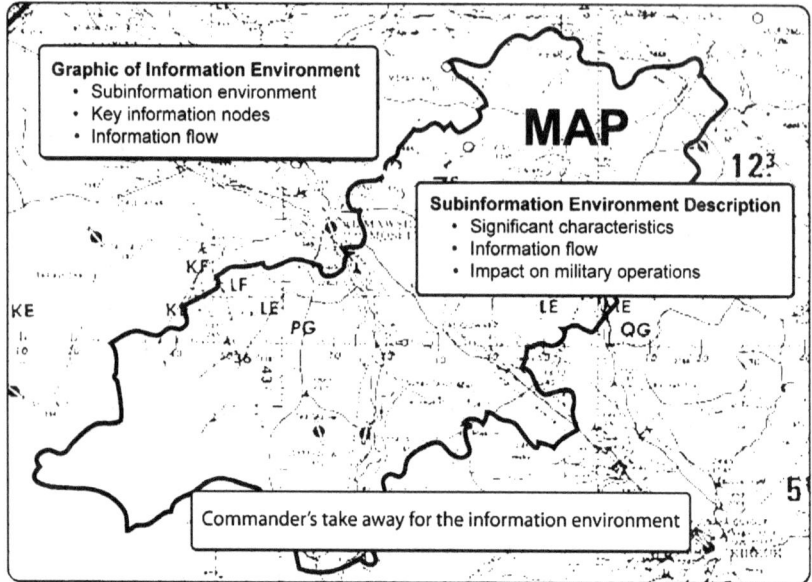

Figure A-21. Combined information operations overlay template

EVALUATE THE THREAT—CENTERS OF GRAVITY

A-42. The purpose of performing a threat COG analysis is to determine and evaluate the enemy's (and others') critical vulnerabilities for exploitation. Because this tool is used to evaluate the threat, the appropriate time to perform this analysis is during step 3 (evaluate the threat) of IPOE. The results of COG analysis are later used during COA development to exploit identified vulnerabilities. Chapter 5, Templating Using Center-of-Gravity Analysis, page 5-7, provides additional information.

A-43. The COG analysis of the threat should be conducted by the G-2. The IO staff will provide input to the COG analysis and use it to determine what aspects of the threat IO should engage. Figure A-23, page A-20, provides a graphical depiction of where the COG analysis falls into the military decisionmaking process.

EVALUATE THE THREAT—TEMPLATING

A-44. More formal modeling produces templates that portray the normal or doctrinal (historical) composition and organization of the adversary's information system and its assets. The result should identify adversary capabilities and vulnerabilities under ideal conditions in the information environment. Templates will vary widely by operation—the examples provided are illustrative only.

Decisionmaking Template

A-45. The decisionmaking template identifies who makes decisions. Its purpose is to identify key leaders, organizational structures, linkages and interrelationships, key decisionmakers, and decisionmaking characteristics. Figure A-24, page A-20, provides a sample decisionmaking template.

Planning Aids

Information Infrastructure Template

A-46. The information infrastructure template identifies what assets and means the adversary uses to collect, protect, and project information. The template identifies critical adversary information system nodes, links, and systems (to include those assets capable of impacting the information environment). Figure A-25, page A-21, provides a sample information infrastructure template.

Information Tactics Template

A-47. The information tactics template identifies how the adversary will collect, protect, and project information. It identifies adversary tactics, past use of information, and available assets. Figure A-26, page A-21, provides a sample information tactics template.

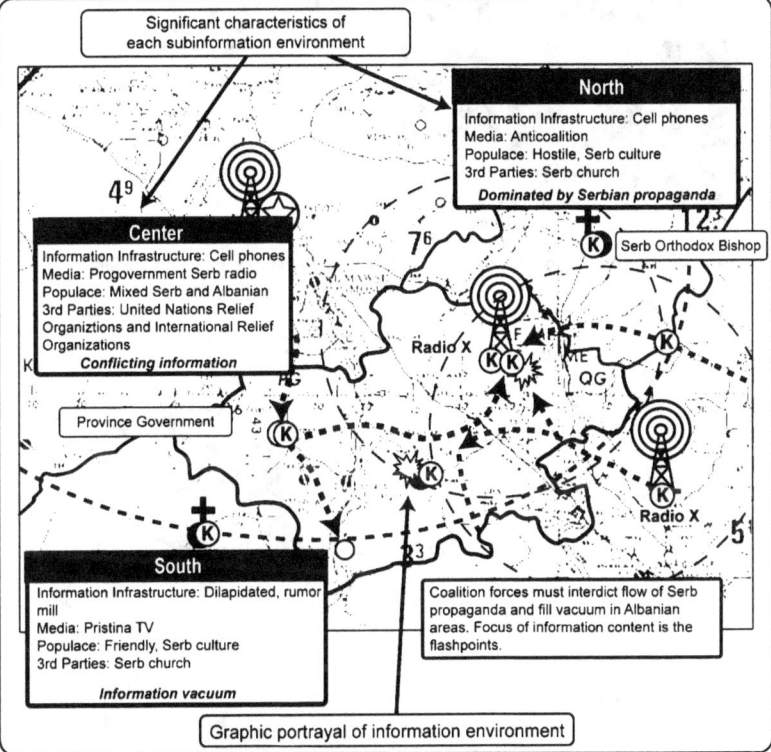

Figure A-22. Example of a combined information operation

Appendix A

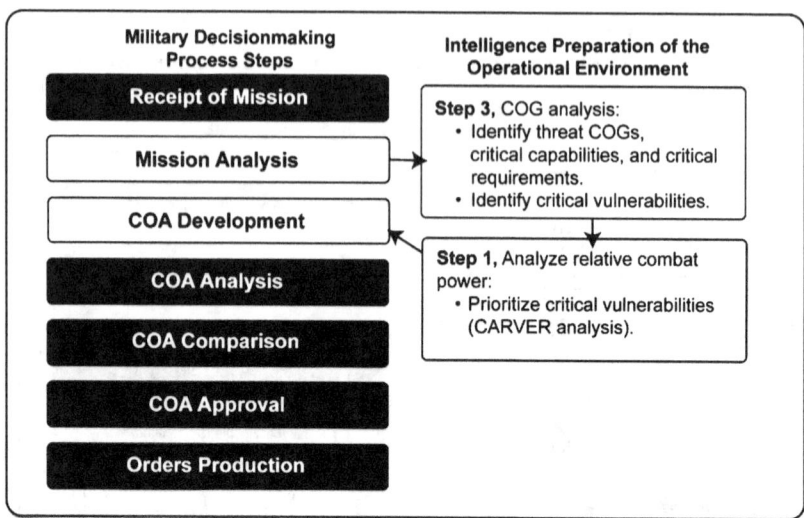

Figure A-23. Relationship between center-of-gravity analysis and the planning process

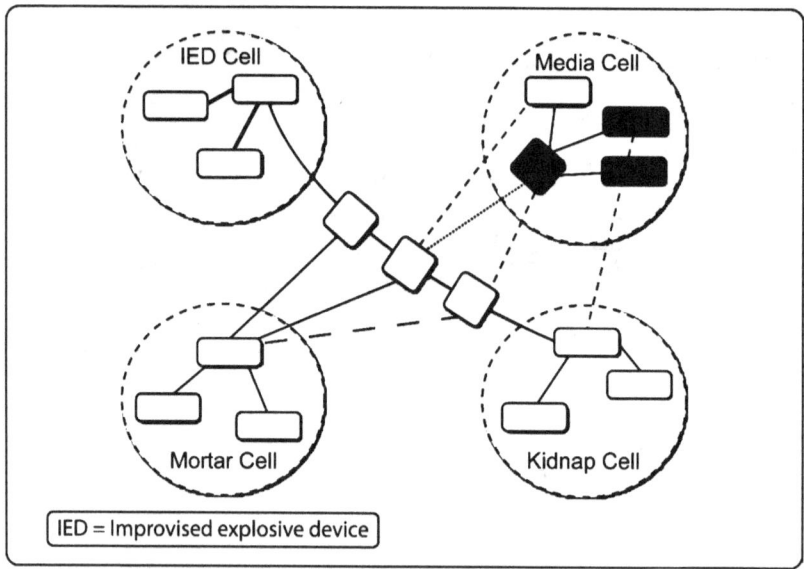

Figure A-24. Decisionmaking template

Planning Aids

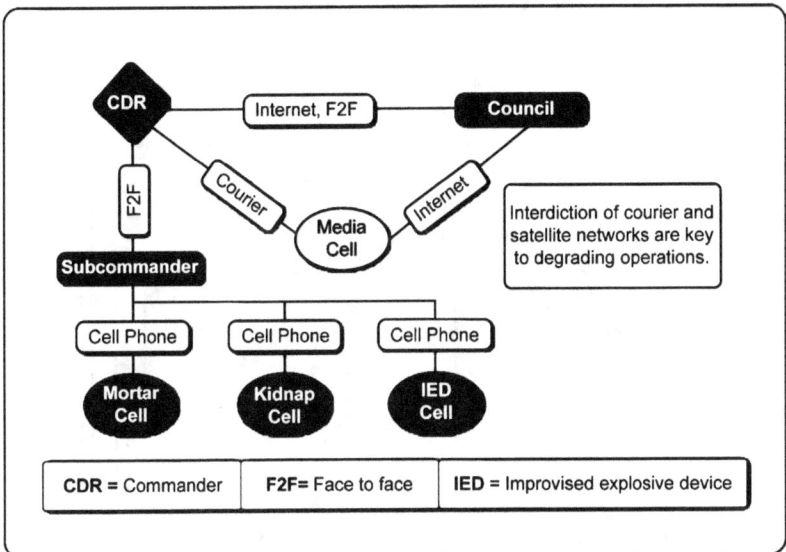

Figure A-25. Information infrastructure template

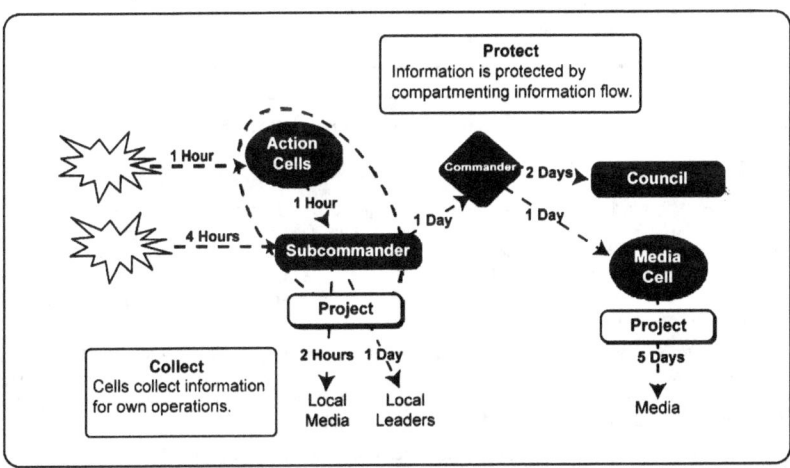

Figure A-26. Information tactics template

Appendix A

DETERMINE THREAT ACTIVITIES IN THE INFORMATION ENVIRONMENT

A-48. The information situation template identifies where, when, and why the adversary will seek information superiority. The result is a concept of operations that describes how the adversary will operate in the information environment. Figure A-27 provides a sample information situation template.

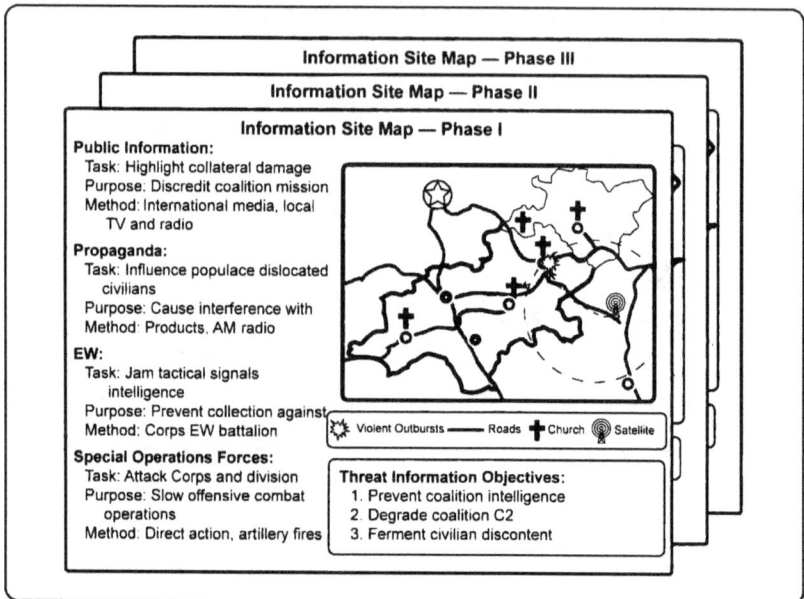

Figure A-27. Information situation template

STAFF ESTIMATE FOR INFORMATION OPERATIONS

A-49. The staff estimate is an assessment of the situation and an analysis of the COAs the commander is considering. The estimate includes an evaluation of how factors in a staff section's functional area influence each COA, and the conclusions and recommendations for each COA. Staff estimates are normally text documents, but may be formatted as maps, graphics, or charts. The estimates are as comprehensive as possible, yet not overly time-consuming to develop. They are developed as part of the planning process, and updated as the operation progresses.

A-50. The staff estimate for IO is an estimate tailored to the specific needs of the IO staff. It assesses the situation in the information environment and analyzes the best way to achieve information superiority. Leaders focus on the information environment and the use of information by enemy and friendly forces. When possible, graphics are added to illustrate the less-tangible aspects of IO.

A-51. The written estimate is a six-paragraph document. The first two paragraphs are necessary for all plans. The other paragraphs can be truncated when time is short. Figure A-28, page A-23, depicts a format for a written IO estimate.

Planning Aids

1. **Mission.** Define information superiority for the mission (Step 1).
2. **Situation and considerations.**
a. Characteristics of the information environment (Step 2).
b. Enemy forces (Step 3).
c. Friendly forces (Step 4).
d. Assumptions (Step 5).
3. **COAs.** List options for achieving information superiority.
4. **Analysis.** Estimate likelihood of accomplishing IO objectives given available time and capabilities.
5. **Comparison.** Compare COAs using evaluation criteria.
6. **Recommendations and conclusions.**
7. **Recommended COA based on which is most supportable by IO.**

Figure A-28. Example of an information operations estimate format

MISSION

A-52. The IO mission describes the operational advantage that IO achieves in support to the unit's mission.

Characteristics of the Information Environment

A-53. The characteristics paragraph describes the significant characteristics of the information environment in terms of the physical, information, and cognitive dimensions. The following characteristics should be considered:
- Terrain.
- Civilian information infrastructure.
- Media.
- Civilian population.
- Third-party organizations.

A-54. The character of each subinformation environment in the AO is reviewed to determine whether it favors friendly or adversary forces. Leaders identify information nodes in each subinformation environment (that is, places, persons, or infrastructure that shape information content and flow by creating or transmitting information).

Enemy Forces

A-55. The enemy forces paragraph describes how, when, where, and why the enemy force operates in the information environment. It identifies enemy capabilities and vulnerabilities in the information environment in terms of information—collection, protection, projection.

Friendly Forces

A-56. The friendly forces paragraph describes friendly-force capabilities to operate in the information environment. It identifies friendly vulnerabilities to enemy and third-party actions in the information environment.

Assumptions

A-57. The assumptions paragraph lists the assumptions essential for planning, execution, and assessment of the information operation. It is organized by—
- Information environment (information content and flow).
- Adversary capabilities and vulnerabilities in the information environment.
- Friendly capabilities and vulnerabilities in the information environment.

Appendix A

GRAPHIC INFORMATION OPERATIONS ESTIMATE

A-58. A graphic IO estimate (Figure A-29) contains the same basic information as a written estimate. It includes—
- Information superiority for the mission.
- Characteristics of the information environment (subinformation environment nodes).
- Enemy vulnerabilities and capabilities in the information environment.
- Friendly-force capabilities and vulnerabilities in the information environment.

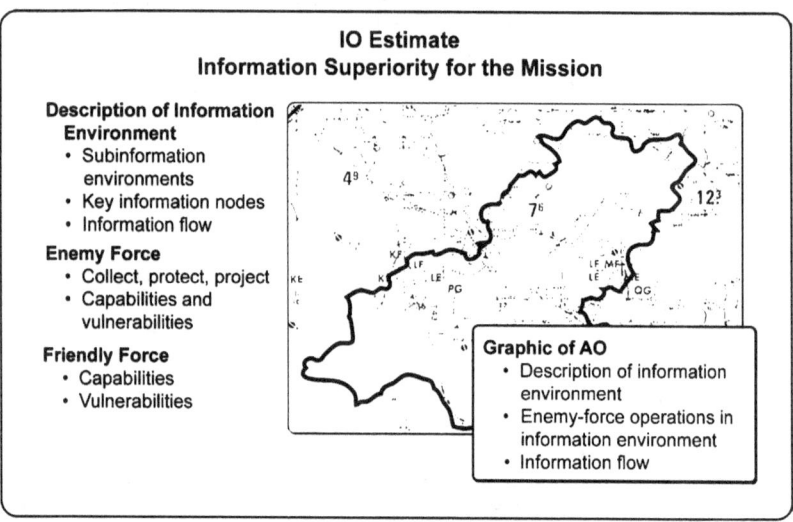

Figure A-29. Graphic information operations estimate

INFORMATION OPERATIONS ANNEX

A-59. Plans and orders are as detailed as time permits. Depending on the command and mission, these documents can be a series of overlays with written comments or they can be voluminous documents of hundreds of pages. Whatever the format, an order must be clear, concise, timely, and useful to the implementing commands and units. The IO annex describes the information operation as a whole and how IO forces will gain information superiority in support of the scheme of maneuver. This approach places less emphasis on individual IO assets and capabilities and more on the aggregate IO effects needed to achieve information superiority. The IO staff must be careful to not let the requirement to develop and explain the IO element contribution to the operation overwhelm the primary purposes of the IO annex, which are to—
- Provide operational details on the information operation.
- Focus element and unit tasks on achieving specific effects in the information environment.
- Provide the information needed to assess the information operation.

A-60. There are two basic formats for an annex:
- *Five-paragraph IO annex.* The five-paragraph annex (Figure 3-12, pages 3-16 and 3-17) is used when time is available and/or when directed by the G-3 or unit SOP.
- *Matrix IO annex.* The matrix annex (Figure 3-13, page 3-17) is used when time is available or when directed by the G-3 or unit SOP.

A-61. Figure A-30 provides an example of an execution matrix.

Example Information Operations Execution Matrix				
Tasked Unit or System	Phase I	Phase II	Phase III	Phase IV

Example Information Operations Execution Matrix				
Element/ Capability	Phase I	Phase II	Phase III	Phase IV
EW				
MISO				
OPSEC				
CNO				
MILDEC				
CMO				
PA				

EXAMPLE

Figure A-30. Example of an execution matrix

INFORMATION OPERATIONS CONCEPT OF SUPPORT

A-62. The IO concept of support is a word picture that explains execution of the information operation from beginning to end and how the capabilities will be employed to gain information superiority. This requires defining information superiority for the operation. A well-written concept is concise and understandable. Although there is no doctrinally prescribed formula for an IO concept of support, planners should consider including the following:

- *Commander's intent for IO.* Explain what the commander wants IO to do to the enemy or the information environment.
- *Information superiority.* Explain specifically what information superiority is within the context of the operational situation and the mission. Include the specific time and location information superiority will be achieved.
- *General scheme for IO.* Use doctrinal concepts and terms to explain how the IO objectives will be achieved, who will perform IO (that is, the tasked units), and the sequencing of key tasks. Relate the key tasks to the achievement of information superiority.
- *Priority of support.* Designate which subordinate unit or element has the priority of IO assets and capabilities.
- *Restrictions on the employment of IO.* List prohibited and directed actions that affect the employment of IO.

This page intentionally left blank.

Appendix B
Tactical Deception Aid

Tactical MILDEC is deception planned and executed by and in support of tactical commanders to result in adversary actions that are favorable to the originator's objectives and operations. The purpose of tactical deception is to mislead or confuse the enemy decisionmaker by distorting, concealing, or falsifying indicators of friendly intentions, capabilities, or dispositions. Figure B-1 provides an overview of the deception planning process. Figure B-2, page B-2, provides a deception estimate format.

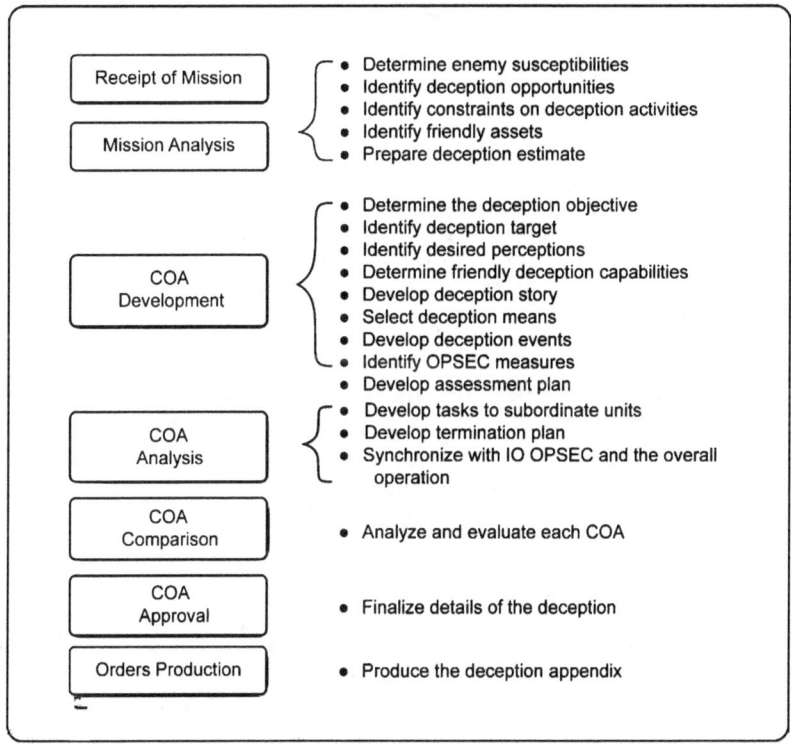

Figure B-1. Deception planning process overview

Appendix B

Mission	• Restated mission of the command. • Deception objective. Identify the purpose of the deception.
Situation and COA	• Summarize the situation in terms of characteristics of the AO, enemy situation, friendly situation, assumptions. • Identify friendly deception COA.
Analysis of Deception COAs	• For each deception COA, list the target-vulnerability analysis, desired perceptions, deception story, means, events, risk analysis, and probability-of-success assessment.
Comparison of Deception COAs	• Compare each deception COA in terms of costs and benefits, operational risks, comparative strengths, weaknesses, and probabilities of success.
Recommendation	• Recommend a deception COA.

Figure B-2. Deception estimate format

DESIRED PERCEPTIONS

B-1. Desired perceptions are those thoughts the target audience must process to believe the planned deception story. The formation of the target audience's perceptions is largely based on the means and events used to portray the deception story.

MEANS

B-2. Considerations for selecting deception means include the following:
- What collection systems or mechanisms does the target audience use?
- How much credibility does the target audience place on information from each conduit?
- What kind of information can be conveyed through each of the means?
- When is each means available to transmit information?
- What filters affect information as it moves through the means?
- How long will it take the information to reach the target audience?

EVENTS

B-3. The deception story is portrayed to the target audience through deception events conducted by friendly forces. These are pieces of a puzzle that the target audience assembles over time. The puzzle itself is the deception story, the pieces are the deceptive events seen by the target audience via the means. Events must be observed and accepted as reality by the target audience. The two types of deception events include—
- Those necessary for the formation of desired perceptions (required events).
- The supporting events that complement or reinforce the desired perceptions.

ASSESSMENT PLAN

B-4. The two primary forms of feedback in deception operations are—
- *Indicator feedback.* This feedback is information that indicates whether and how the deception story is reaching the deception target audience. This feedback is useful for the timing and sequencing of executions. (It answers the question: "Is the target audience receiving the deception story as planned?")
- *Perception feedback.* This feedback is information that shows whether the target audience is forming the desired perceptions and is acting (or is likely to act) in accordance with the deception objective. (It answers the question: "Is the target audience acting in accordance with the deception objective?")

B-5. At the tactical level, the pace of operations and limited number of collection assets may reduce the practicality and utility of feedback. For this reason, at the lowest levels of command, tactical deception operations must not depend on feedback for successful execution.

DECEPTION TECHNIQUES

B-6. Tactical deceptions often contain one or more of the following techniques:
- *Feint.* A feint is a limited operation to deceive the enemy of the location or time of the decisive operation. Forces seek direct fire contact with the enemy but avoid decisive engagement. Feints usually occur before or during the main operation. Multiple feints may be needed to portray the deception story. The objective of a feint is to cause the enemy to misemploy forces.
- *Demonstration.* Demonstrations are shows of force to deceive the enemy as to the location or time of the decisive operation. They are similar to feints, except no contact is made with the enemy. The objective is to delude the enemy into an unfavorable COA. Demonstrations are useful when time and distance factors make the lack of contact realistic.
- *Ruse.* A ruse is a deliberate exposure of false information to enemy collection means.
- *Display.* A static display of an activity, force, or equipment is intended to deceive enemy observation. Displays project the appearance of objects that do not exist or appear to be something else. Observables include the use of heat, smoke, electronic emissions, false tracks, and fake command posts.

DECEPTION TACTICS

B-7. The two types of deception tactics are ambiguity-increasing deception and ambiguity-reducing deception. The following paragraphs discussed these tactics.

AMBIGUITY-INCREASING DECEPTION

B-8. Ambiguity-increasing deception increases decisionmaker uncertainty about key information needed to make decisions. It can be used to delay a specific decision or reduce the quality of a decision. Ambiguity-increasing deception—
- Presents conflicting elements of information.
- Overloads enemy intelligence-collection and analytical capabilities.
- Confuses enemy expectations about friendly-force size, activity, location, unit, time, equipment, intent, or mission.

AMBIGUITY-DECREASING DECEPTION

B-9. Ambiguity-decreasing deception provides the decisionmaker with the illusion of reduced uncertainty and risk. It can be used to elicit specific behavior that can be exploited by friendly forces and to provide cover for friendly actions. Ambiguity-decreasing deception—
- Reinforces the enemy's preconceived beliefs.
- Draws enemy attention from one set of activities to another.
- Creates the illusion of strength where weakness exists.
- Creates the illusion of weakness where strength exists.
- Accustoms the enemy to particular patterns of activity that are exploitable later.

MILITARY INFORMATION SUPPORT OPERATIONS IN SUPPORT OF DECEPTION OPERATIONS

B-10. At the tactical level, MISO are a primary deception capability. MIS units may conduct tactical deception by using sonic deception (that is, loudspeakers) for protection and in support of direct action missions. MIS forces may also develop, modify, and disseminate print and audiovisual products to support a deception operation.

DECEPTION IN SUPPORT OF COMBAT OPERATIONS

B-11. In support of combat operations, deception can preserve friendly forces and equipment from destruction, gain time, or minimize an enemy's advantage. A deception is most effective if the friendly force has more COAs available than the enemy has forces to cover in strength. The purpose of deception in combat operations is to create an operational advantage through surprise (that is, specific time, place, method, and scope of an attack). Possible objectives include the following:

- Delay or prevent the enemy's action or counteraction.
- Cause the enemy to misdirect assets.
- Cause the enemy to employ forces in ways that makes them vulnerable to the friendly COA.
- Cause the enemy to reveal strengths, dispositions, and intentions.
- Cause the enemy to waste combat power with inappropriate or delayed actions.

DECEPTION IN SUPPORT OF STABILITY OPERATIONS

B-12. Deception is appropriate during stability and support operations when transparency of operations is a likely requirement. Deception may serve to protect U.S. Soldiers, mask operational intentions, and deter adversary factions. The purpose of deception during such operations is to degrade adversary attempts to disrupt peace. Possible objectives include the following:

- Cause hostile forces to not attack friendly forces (protection).
- Deter factional violence.

B-13. Political objectives may override military considerations, to include the use of deception. Participation of multinational forces also may restrict the utility and use of deception.

DECEPTION IN SUPPORT OF OPERATIONS SECURITY

B-14. Deception in support of OPSEC increases the likely detection of indicators that the enemy can observe to derive an incorrect conclusion. OPSEC hides real indicators, whereas deception shows fake indicators. Observables are presented to distract enemy intelligence collection away from (or provide cover for) real friendly operations and activities.

DECEPTION IN SUPPORT OF COUNTERINSURGENCY

B-15. During counterinsurgency missions, in-depth human-factors analysis of deception target audiences may not be possible. In lieu of a human-factors analysis of the target audiences, planners can use profiles of cell leaders or security organizers. Counterdeception is important, as insurgent and guerrilla warfare theory emphasizes the use of deception to accomplish goals.

Appendix C
Tactical Operations Security Aid

OPSEC is a universal IO capability. It should be included in all plans, operations, and activities. The OPSEC process is a framework to systematically identify, analyze, and protect information. The goal of OPSEC, in conjunction with unit security programs, is to achieve essential secrecy. The OPSEC process should be integrated into the military decisionmaking process. It uses the steps indicated in Figure C-1, but does not have to follow them in a particular sequence.

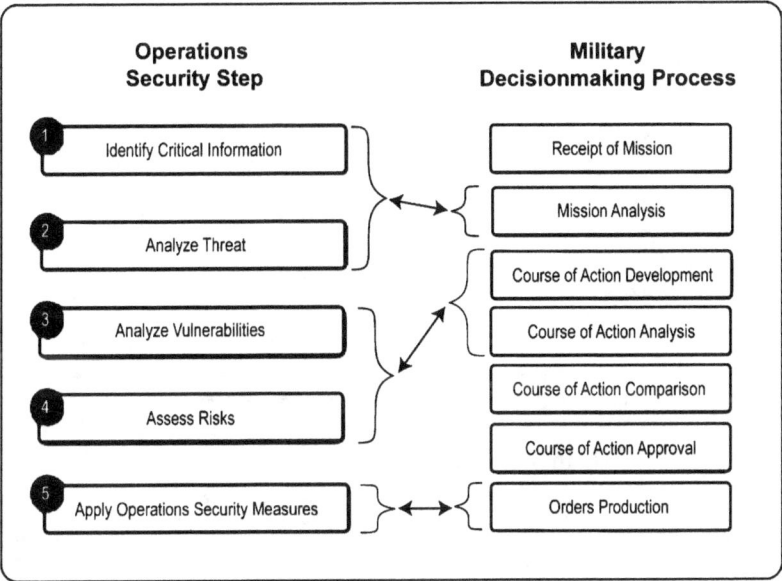

Figure C-1. Operations security and the planning process

IDENTIFY CRITICAL INFORMATION

C-1. Planners determine what information must be protected (that is, a list of EEFI). Sources of EEFI include the—
- Higher HQ plans and operation orders.
- Commander's guidance.
- Current unit EEFI.

Appendix C

C-2. EEFI focus on friendly-force intentions (time and place of units and operations), capabilities, and vulnerabilities (strength, technologies, and tactics). EEFI are different for every operation. Leaders must avoid the "cookie-cutter" approach and should continually develop new EEFI or refine old EEFI. The OPSEC working group can be used to take advantage of subject-matter experts (for example, aviation and communications). Leaders identify the length of time each EEFI must be protected (not all information needs protection for the duration of the operation). EEFI are prioritized and kept to a manageable number (perhaps five).

ANALYZE THE THREAT

C-3. The threat to EEFI is the sum of enemy information needs and enemy collection capabilities. A CI template (Figure C-2) is a useful tool to depict enemy collection capabilities. It shows when and where the EEFI are vulnerable to enemy collection.

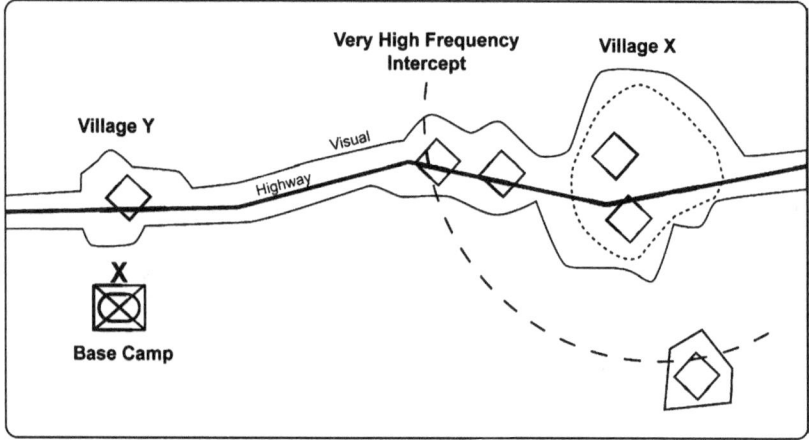

Figure C-2. Example of a counterintelligence template

ANALYZE VULNERABILITIES

C-4. Leaders identify each EEFI's vulnerability to enemy intelligence collection (that is, OPSEC vulnerability). The OPSEC vulnerability is a result of the OPSEC indicator and enemy collection capabilities. OPSEC vulnerabilities are detectable indicators of EEFI. OPSEC indicators become OPSEC vulnerabilities if they can be observed, analyzed, and acted upon by the enemy. To determine OPSEC vulnerabilities, leaders—

- *Identify OPSEC indicators.* Leaders determine what detectable actions and OSINT can be interpreted or pieced together by the enemy to derive EEFI.
- *Compare OPSEC indicators to enemy collection capabilities.* Leaders determine which indicators can be observed, analyzed, and acted upon by the enemy.

ASSESS RISK

C-5. Leaders develop measures to protect OPSEC vulnerabilities by conducting risk assessments for each vulnerability and then selecting one or more OPSEC measure for each vulnerability. There are three types of OPSEC measures:

- *Action controls.* The controls change unit procedures, activities, and actions (for example, randomized routine activities, avoiding repetitive TTP).

Tactical Operations Security Aid

- *Countermeasures.* These measures disrupt enemy information-gathering and targeting (for example jamming [EW], physical attack, and camouflage and concealment).
- *Counteranalysis.* This action deceives the enemy by providing false indicators (for example, decoys and deception in support of OPSEC).

C-6. Once leaders decide which OPSEC measures to implement, they must check that OPSEC measures do not create new vulnerabilities. Leaders must balance OPSEC measures with operational effectiveness.

APPLY OPERATION SECURITY MEASURES

C-7. Leaders apply OPSEC tasks to units and staff as follows:
- Rewrite approved OPSEC measures as tasks.
- Assign responsibility and coordinate OPSEC tasks with units and staff.
- Coordinate OPSEC measures with MILDEC, PA, and COMCAM to prevent compromise of EEFI.
- Integrate OPSEC tasks with IO.
- Include OPSEC tasks in the operation order/operation plan.
- Adjust OPSEC measures based on adversary reaction to the implemented OPSEC measures.
- Monitor execution.
- Evaluate effectiveness.
- Adjust measures and tasks.
- Coordinate monitoring of OPSEC measures through the G-2 and CI.

This page intentionally left blank.

Appendix D
Media Assessment Aid

Media analysis is a quick and useful technique for evaluating the impact of media coverage on military operations.

IDENTIFY MEDIA SOURCES

D-1. Planners identify media outlets that are critical to mission accomplishment by analyzing the flow of media reporting in the AO and area of interest, selecting those media outlets that have local, regional, or international influence. Media outlets that are overtly biased toward the adversary should not be used. A good sampling of media outlets includes the following:
- *Local media in the AO.* These outlets influence local public opinion.
- *Regional media in countries adjacent to the AO.* These outlets can influence public and political opinion in the AO and area of interest. For example, in Afghanistan it is important to monitor the Pakistani press.
- *International media are the larger media outlets associated with countries outside the AO.* Typically these outlets impact U.S. domestic, coalition partner, and worldwide public and political opinion (for example, in Canada, the Canadian Broadcasting Corporation; in the United Kingdom, the British Broadcasting Corporation; in Germany, *Der Spiegel;* and in the United States, the Cable News Network, *The Washington Post, Los Angeles Times,* and *The New York Times*).

Note. Home (domestic) media is a primary consideration for the PA staff, but it is not a consideration for military IO.

D-2. The media analysis presented in this aid is a tool that staffs can use to understand and assess the impact of media reporting on friendly and enemy activities in the AO. This type of media analysis helps the staff—
- Maintain situational awareness on media reporting.
- Evaluate the impact of media reporting on the mission.
- Identify adversary information.
- Provide data for assessment.

D-3. Other staff elements may also conduct media analysis to support their functional area:
- The PAO conducts a media content analysis to assess news coverage.
- The intelligence staff may collect and analyze media reporting as part of OSINT.
- MIS forces prepare extensive media assessments and analyses of commercial and government media within their AO, as they seek to leverage more indigenous and credible local media outlets to use for dissemination.

D-4. The IO staff must be prepared to analyze the media, to monitor changes in the information environment, and to counter adversary misinformation and propaganda.

Appendix D

DATA COLLECTION

D-5. Planners systematically monitor media coverage of the command, its mission, and the AO from the sources identified above. Useful sources of media reports and stories include the following:
- *PAO media operations center.* This center provides translations of foreign press coverage in addition to monitoring major English-language media outlets.
- *OSINT media-monitoring sources (contracted by the Department of Defense).*
- *United States Government open-source center.*
- *Internet.*

D-6. Data collection must be continuous and consistent—usually on a daily basis. Several factors can affect collection of data reports. Items that need to be considered include the following:
- English-language media sources are readily available and may skew the collection effort away from local media.
- Translation of local and regional media may cause a lag time of a day or more.
- A database should be created, populated, and maintained to establish a baseline upon which comparisons can be made (for example, media reporting for one month versus another month).

D-7. Other than the PAO's media content analysis, there is no established doctrinal method for analyzing the media. The media analysis process identified in Figure D-1 has been field-tested. It can be modified to fit command and staff needs.

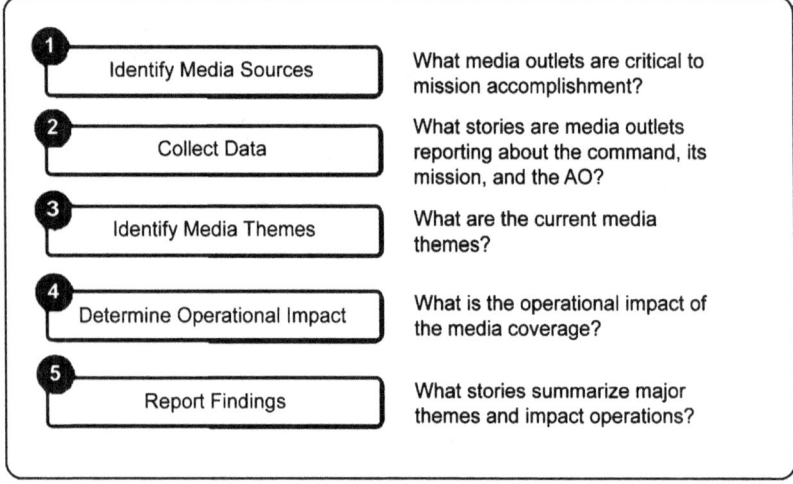

Figure D-1. Media analysis process

IDENTIFY MEDIA THEMES

D-8. Planners analyze and plot the media coverage collected above to identify current media themes (Figure D-2). Planners should do this by—
- Picking out primary themes in media reports and stories.
- Categorizing themes into groups that either support (positive), run counter to the command's objectives (negative), or are neutral to either enemy or friendly forces.
- Identifying ad hoc themes of interest to the command.
- Pairing media sources and themes to the command's objectives or lines of operation.

Objective	Theme (+/–)	Source
Maintain international support for mission	(+) United Kingdom supports troop expansion	British Broadcasting Corporation News and Cable News Network
	(–) U.S. missile strikes kill civilians	British Broadcasting Corporation News and Associated Press
Objective	Theme (+/–)	Objective
Reduce popular support for insurgents	(+) Tribal elders turn in Taliban to Afghan police	Local Armed Forces radio and TV
	(–) Taliban propaganda about U.S. missile strikes	*Dawn* (Pakistani newspaper)

Figure D-2. Media theme assessment diagram

DETERMINE OPERATIONAL IMPACT

D-9. Planners analyze the themes identified above to determine the impact on friendly and enemy operations. Media sources and themes are categorized by echelon (that is, local, regional, and international). For each theme, planners answer the following questions:
- Who is the originating source of the theme:
  - Enemy or hostile forces?
  - Friendly forces?
  - Third-party organization?
  - Embedded media?
- Who is the target audience?
- What is the circulation of theme (most critical for local media)?
- What are the second- and third-order effects?
- Is the event affected by extended media coverage?

D-10. Themes are prioritized within each category based on degree of impact. Planners determine which negative themes pose potential problems for ongoing and future operations and determine which positive themes provide an opportunity for exploitation.

REPORT FINDINGS

D-11. There is no standard method on how to report media findings. The key is to portray media coverage in an easily understood format that can be quickly scanned to see what themes are important. Planners should use color coding to clearly display the impact of each theme: green is positive, red is negative, and blue is neutral. Symbols (for example, +/–, letters, or numbers) should be added so the analysis can be understood if printed in black and white. Planners must resist the temptation to fill the

Appendix D

boxes with headlines rather than themes. Media themes that reflect enemy propaganda should be added, along with an assessment of operational impacts. Planners display a trend analysis to put the current media reporting into a broader context. Figure D-3 provides one technique of reporting findings.

CONSEQUENCE MANAGEMENT

D-12. Military operations can trigger either positive or negative coverage by the media. This coverage may be a situation that must be mitigated to prevent or reduce the impact on the unit mission, or exploited as an opportunity to further the command's objectives. Such a situation is called "media bounce."

D-13. Media bounce refers to the staying power of a story over time. The bounce is usually short, particularly if another newsworthy event occurs. Monitoring media bounce avoids reacting to an event that loses media's attention and may renew negative reporting, thereby aggravating the situation. A consequence-management tracker (Figures D-4 and D-5, page D-5) is a simple decisionmaking aid that tracks subsequent media reporting of an event (bounce) to determine whether subsequent command action is required.

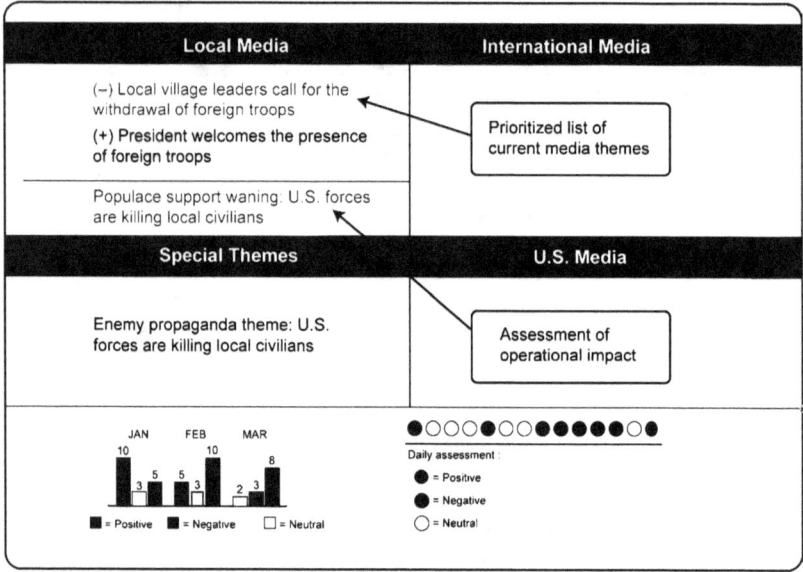

Figure D-3. Sample media report

Media Assessment Aid

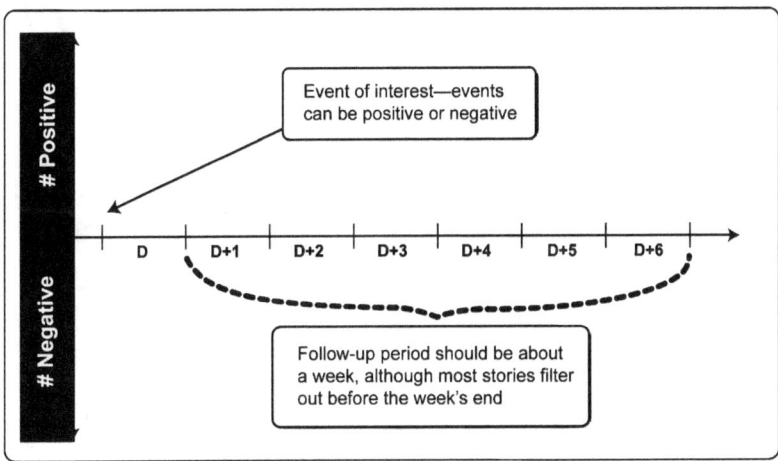

Figure D-4. Consequence-management tracker format

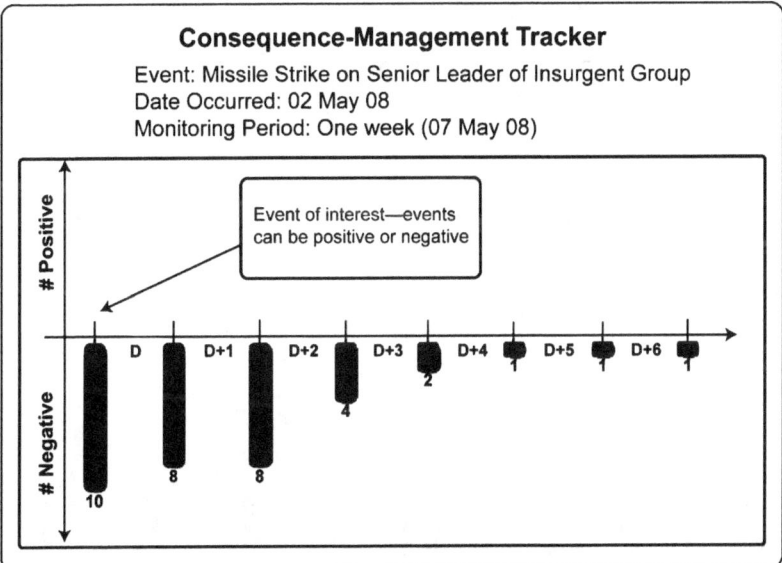

Figure D-5. Example of a consequence-management tracker

This page intentionally left blank.

Appendix E
Conducting Face-to-Face Meetings

This appendix provides a guide for tactical face-to-face engagements by troops at the detachment, company, battalion, and group level when meeting target audiences.

Leaders must always remember they are representing their unit, the command, and, for all intents and purposes, the United States and its allies. Regardless of rank or position, actions taken when in contact with the populace can shape the battlefield to defeat an enemy seeking local sanctuary to conduct attacks on the forces and allies of the United States.

PREPARING FOR A FACE-TO-FACE MEETING

E-1. Leaders can take a number of steps to prepare for face-to-face meetings. These steps include:
- *Research.* Leaders should learn everything they can about the target audience, to include proper name and title, approximate age, family members, ethnicity, language spoken, and the target-audience's relationship to other leaders, friendly forces, third-party organizations, and the adversary. Reliance on intelligence and MISO capabilities provides a strong foundation of information to leaders prior to a face-to-face meeting.
- *Check previous contacts.* Leaders should determine who has had contact with the target audience; when the meeting took place; what was discussed; what promises were made; whether the target audience was deemed truthful, manipulative, or trustworthy; and the groups or individuals to which the target audience is tied. With attached or assigned MIS forces managing the key-leader engagement program, this information should be catalogued and ready for access to guide future contacts.
- *Keep records.* Leaders should take notes during the conversation (either personally or via an aide) referring back to them at the end to capture the essence of the conversation. Notes should be shared with other interested persons, particularly MIS planners.
- *Coordinate.* Face-to-face meetings are coordinated to prevent other friendly forces from sending mixed messages to the target audience.
- *Set a time limit.* Leaders determine how long the meeting should be, staying as close to the timeline as possible while exploiting any available opportunities.
- *Consider perceptions.* Many factors affect target-audience perceptions (for example, uniforms, long versus short guns, large versus small convoys, aircraft in the area, type and size of escort, civilians present, and the number of people attending the meeting).
- *Plan for problems.* Leaders establish code words to maintain control of information flow and security. Typical situations where code words may be used include the desire to end the conversation, a noted potential for violence or increased threat, and other possible emergencies.
- *Rehearse.* Leaders practice the discussion with another person through the translator. Comments should be solicited from anyone having experience with the target audience.
- *Plan the rest of the operation.* Leaders should plan face-to-face meetings with the same intensity and focus as they would plan combat operations. Special items of note include—
    - Translator integration.
    - Movement (ingress and egress).
    - Security (both sides know of planned meetings, so units must anticipate compromise).
    - Contingency and emergency situations and covert danger signals.

Appendix E

CONDUCTING A FACE-TO-FACE MEETING

E-2. When conducting face-to-face meetings, leaders should—

- *Bring past notes or previous reports for reference.* This demonstrates interest as well as directs the conversation into favorable areas.
- *Perform introductions.* Leaders introduce everyone in their party and record the names and positions of everyone outside their party who is attending. Collecting information is a key goal of each meeting.
- *Take photographs.* After asking (and receiving) permission, leaders should take photographs of the target audience.
- *Be sincere.* Leaders may wish to apologize in advance for any cultural mistakes made, reassuring the target audience that no offense is intended. Leaders may ask the target audience to point out errors as a learning tool. As the face-to-face meeting ends, the leader may wish to ask what cultural mistakes were made and thank the target audience for helping the leader to learn the local culture.
- *Avoid restricted topics.* Leaders should not discuss sensitive issues such as religion or other societal practices.
- *Compare notes.* Immediately after the meeting, the attendees should discuss what was observed to ensure an accurate understanding of what occurred.
- *Avoid false assumptions.* Leaders should never assume the target audience does not understand English.

CHANCE ENCOUNTERS AND CONTACTS

E-3. A chance encounter or contact with the target audience occurs most often during patrols at the squad, platoon, and company levels. The leader of the unit should conduct the face-to-face meeting based upon a preplanned battle drill. Items to consider during chance encounters include the following:

- *Maintain security.* The leader of the patrol should preserve security of the communicator and the target audience.
- *Maintain schedule.* The leader of the patrol should limit the length of the face-to-face meeting by establishing a code word for when it is time to end the meeting.
- *Identify the local leader.* The leader of the patrol should ask who is in charge and talk to him only. The patrol should not distribute anything to the populace without the local leader's permission.
- *Be fair and firm.* The leader of the patrol should stay in charge and be respectful, not rude.
- *Be selective.* The leader of the patrol should select a maximum of one or two people to talk with.
- *Take notes.* The patrol should get names of all people contacted, approximate ages, hometown, business or activity, subjects covered, demeanor towards friendly forces, and any particular concerns of the target audience. This information should be shared with the intelligence and MISO capabilities at the first opportunity after the contact.
- *Be prudent.* The patrol should not make promises that cannot be kept.
- *Establish rapport.* The leader of the patrol should offer the target audience refreshments (such as a bottle of water) and move to a comfortable location. Sit, if possible.
- *Focus.* The patrol should stay on message by knowing what messages the command is focusing on in specific AOs and during specific time periods.
- *Reinforce the message.* The patrol should use any applicable/available printed products (handbills, pamphlets, or posters) to reinforce the verbal message and request a formal follow-up engagement, if deemed necessary, based on the issues discussed.
- *Report.* The patrol should report all contacts with local leaders up the chain of command to ensure that an accurate picture of the situation is developed.

Appendix F
How to Use Translators

The following information is specifically applicable to category I (local hire, uncleared) translators. However, some aspects of the information are applicable to category II (cleared for SECRET) and category III (cleared for TOP SECRET) translators.

Leaders must remember that the translator is their voice and their representative to the community. The translator will be seen as a representative of the command, of the Army, and of the United States. As such, leaders must monitor and keep all aspects of their behavior professional and ethical regardless of their nationality or ethnicity.

GENERAL GUIDELINES

F-1. Leaders should insist translators—
- Speak in first person.
- Remain nearby when the leader is speaking.
- Carry a notepad and take notes, as needed.
- Project clearly and mirror the vocal stress and overall tone of the leader.

F-2. Good leaders know their translators. The lives of Soldiers may be in the translator's hands, so it is critical to know the translator's strengths and weaknesses. Translators should be treated as part of the unit. The better the translator is integrated into the unit, the better the translator's performance.

F-3. Translators should be used for translation duties only. Using them for other activities may violate their contract. An example of misemployment is using a translator to run errands in town. However, sending the translator to town to coordinate a meeting for U.S. officials is allowed.

F-4. The translator is the leader's voice and, as such, may be subject to physical harm because of the messages delivered. Translators should be offered physical protection. If the translator is allowed to carry a weapon, the unit must ensure that he can handle it in a safe manner. Range familiarization/qualification (as well as knowledge of movement techniques and chemical, biological, radiological, and nuclear equipment) is highly recommended.

F-5. Translators should be allowed rest periods to collect their thoughts and catch their breath. Meal meetings are especially challenging for a translator. Leaders should ensure the translator is allowed to eat during or after the meeting.

F-6. Translators should be dressed like the troops they are supporting so they can be readily identified as a friendly in a combat situation to preclude fratricide. Uniform accessories (such as wet weather gear, body armor, and glint tape) that are common on Soldiers' uniforms should be made available to the translator.

REHEARSING WITH A TRANSLATOR

F-7. Leaders must check the translator to verify their abilities. To ensure accuracy and security, periodically record your translator, both with and without his knowledge, for quality checks by higher HQ. If operational details are briefed to the translator during the mission rehearsal, units may consider having the translator remain on the base camp until execution. Also ensure the translator does not have a cellular telephone or other communication device.

F-8. Leaders must rehearse conversations, particularly when dealing with complex, new, or sensitive issues. A rehearsal will help define words the translator may not know and ensure the translator understands the overall message to be conveyed. Leaders provide feedback to the translator and make corrections as needed. Leaders must keep in mind that if the translator performs poorly, it affects the target audience's perception of the unit.

WORKING WITH A TRANSLATOR

F-9. Leaders must always maintain eye contact with the person they are speaking with and not the translator. The target audience should be observed for gestures, posture, and body language.

F-10. Leaders using translators should speak in short clips and not recite long paragraphs. The goal is to make the target audience feel like they are conversing and not being lectured. One to two sentences at a time is a good rule. Acronyms, slang, and idioms should be avoided.

F-11. For simple ideas or routine information, some leaders feel confident that the translator is capable of delivering the intended message. This technique works best if the leader introduces the topic and then expresses confidence in the translator's ability to speak on the leader's behalf. Conversations should be ended with closing comments and an opportunity for questions.

BATTLE DRILLS AND STANDING OPERATING PROCEDURES

F-12. There are some situations when a leader may want to establish battle drills or SOPs that address information the translator will convey to the local populace. Some possible situations and the type of information that the translators should be prepared to provide are—
- *Vehicle checkpoints.* Common concerns include questions about where the vehicle occupants are going or if they are armed, and instructions on where vehicle occupants should stand or what they should do during the search.
- *Cordon and search.* Common concerns include an explanation of what military forces are doing when they question residents about any weapons or suspicious activities.
- *Detention of a person.* Common concerns include an explanation telling why the person was detained, how the detention process works, how the detainee's family can reach the detainee, and how friendly forces humanely treat detainees.

Glossary

SECTION I – ACRONYMS AND ABBREVIATIONS

ADP	Army Doctrine Publication
AO	area of operations
C2	command and control
CA	Civil Affairs
CARVER	criticality, accessibility, recuperability, vulnerability, effect, and recognizability
CCIR	commander's critical information requirement
CI	counterintelligence
CIO	combined information overlay
CJSOTF-AP	combined joint special operations task force—Arabian Peninsula
CMO	civil-military operations
CMOC	civil-military operations center
CNA	computer network attack
CND	computer network defense
CNO	computer network operations
COA	course of action
COG	center of gravity
COMCAM	combat camera
DA	Department of the Army
DSPD	defense support to public diplomacy
EA	electronic attack
EEFI	essential elements of friendly information
EIOT	essential information operations task
EMS	electromagnetic spectrum
EW	electronic warfare
FFIR	friendly-force information requirement
FID	foreign internal defense
FM	field manual
G-2	assistant chief of staff, intelligence staff section
G-3	assistant chief of staff, operations staff section
G-5	assistant chief of staff, plans staff section
G-6	assistant chief of staff, command, control, communications, and computer systems staff section
G-7	assistant chief of staff, information operations
HN	host nation
HQ	headquarters
HUMINT	human intelligence

Glossary

IA	information assurance
IO	information operations
IOWG	information operations working group
IPOE	intelligence preparation of the operational environment
J-2	intelligence directorate of a joint staff
J-3	operations directorate of a joint staff
J-5	plans directorate of a joint staff
J-6	command, control, communications, and computer systems directorate of a joint staff
JCCC	joint combat camera center
JP	joint publication
JSOTF	joint special operations task force
MASINT	measurement and signature intelligence
MILDEC	military deception
MIS	Military Information Support
MISO	Military Information Support operations
MNF	multinational force-Iraq
MOE	measure of effectiveness
MOP	measure of performance
OPSEC	operations security
OSINT	open-source intelligence
PA	public affairs
PAO	public affairs officer
PIR	priority intelligence requirement
S-2	intelligence staff officer
S-3	operations staff officer
S-6	command, control, communications, and computer systems staff officer
S-7	information operations staff officer
SF	Special Forces
SFG(A)	Special Forces group (Airborne)
SFODA	Special Forces operational detachment A
SFODB	Special Forces operational detachment B
SIGINT	signals intelligence
SOP	standing operating procedures
SOTF	special operations task force
TC	training circular
TTP	tactics, techniques, and procedures
TV	television
U.S.	United States
USAJFKSWCS	United States Army John F. Kennedy Special Warfare Center and School
USD(P)	Under Secretary of Defense for Policy

USG United States Government

SECTION II – TERMS

information environment
The aggregate of individuals, organizations, and systems that collect, process, disseminate, or act on information. (JP 1-02)

information operations
The integrated employment, during military operations, of information-related capabilities in concert with other lines of operation to influence, disrupt, corrupt, or usurp the decisionmaking of adversaries and potential adversaries while protecting our own. Also called **IO**. (JP 1-02)

information superiority
The operational advantage derived from the ability to collect, process, and disseminate an uninterrupted flow of information while exploiting or denying an adversary's ability to do the same. (JP 1-02)

This page intentionally left blank.

References

SOURCES USED
These are the sources quoted or paraphrased in this publication.

ARMY PUBLICATIONS
ADP 3-0, *Unified Land Operations*, 10 October 2011.
ADP 5-0, *The Operations Process*, 17 May 2012.
FM 3-05.301, *Psychological Operations Process, Tactics, Techniques, and Procedures*, 30 August 2007.
FM 3-13, *Inform and Influence Activities*, 25 January 2013.
FM 3-53, *Military Information Support Operations*, 4 January 2013.

ARMY FORMS
DA Forms are available on the Army Publishing Directorate web site (www.apd.army.mil).
DA Form 2028 *(Recommended Changes to Publications and Blank Forms)*.

JOINT PUBLICATIONS
JP 1-02, *Department of Defense Dictionary of Military and Associated Terms*, 8 November 2010.
JP 3-13, *Information Operations*, 13 February 2006.

This page intentionally left blank.

Index

C

chance encounters and contacts, 2-20

civil-military operations (CMO), 1-2, 2-1, 2-19, 2-30, 2-32, 2-39 and 2-40, 3-1, 3-7, 3-13 and 3-14, 3-18 and 3-19, 4-7

combat camera (COMCAM), 1-2, 1-5, 2-1, 2-17 through 2-19, 2-30 and 2-31, 2-39 and 2-40, 3-1 and 3-2, 3-5 and 3-6, 3-18, 4-7, A-2, C-3,

combined information overlay (CIO), 3-2, 5-3, 5-5 and 5-6, 5-9, A-15, A-17

command information message, 2-20

commander's critical information requirement (CCIR), 3-6, 4-1, A-11 and A-12, A-14 and A-15

computer network attack, 2-16 and 2-17, 2-33, 2-35, 2-38,

computer network defense (CND), 2-16, 2-32 through 2-38

computer network exploitation, 2-16 and 2-17

computer network operations (CNO), vi, 1-2, 1-5, 2-1, 2-16 and 2-17, 3-1, 3-5, 3-13, 3-18

concept of support statements and sketches, 3-9

consequence management, 2-31, 4-3

constraints on information operations, 3-5

countering adversary IO, 2-28

counterintelligence (CI), 1-2, 2-2, 2-4 and 2-5, 2-32 through 2-38, 3-13, C-2 and C-3

crisis-action team, 4-7

critical events, 4-4, A-5

critical information, iv, 2-2 through 2-5, 2-8, 2-19, 3-2, 3-6, 3-8, 3-12, 5-7, A-10, A-15,

criticality, accessibility, recuperability, vulnerability, effect, and recognizability (CARVER), iv, 5-7, 5-8

D

deception in support of OPSEC, 2-4, 2-6, C-3

defense support to public diplomacy (DSPD), 1-2, 2-31, 2-39 and 2-40

demining operations, 2-13

E

effects against the adversary, 3-10

effects for information operations, 3-11

effects to protect friendly forces, 3-10

effects to shape the environment, 3-11

electromagnetic spectrum (EMS), 2-1, 2-15 and 2-16

electronic attack (EA), 2-15 and 2-16, 2-32 through 2-37, 3-5 and 3-6, 3-12, 3-17

electronic warfare (EW), vi, 1-2, 1-5, 2-1, 2-4, 2-15 and 2-16, 2-32 through 2-38, 3-5, 3-12 through 3-14, 3-17 through 3-19, 4-2, 4-7, A-1, C-3

essential elements of friendly information (EEFI), 2-3 through 2-5, 2-19, 2-32, 2-35, 2-39, 3-2, 3-6 and 3-7, 4-4 and 4-5, A-11 and A-12, A-14 and A-15, C-1 through C-3

essential information operations task (EIOT), 3-12

essential tasks for information operations, 3-2

EW planner, 1-5

F

face-to-face meeting, 2-12, 2-20, 2-22, E-1 and E-2

force protection officer, 2-5

foreign internal defense (FID), 1-5, 2-18, 2-31, 3-10, 3-19

friendly-force information requirement (FFIR), A-12, A-15

G

graphic information operations estimate, 3-4

H

humanitarian assistance, 1-4, 2-13, 2-30

I

information assurance (IA), 1-2, 2-16, 2-32 through, 2-38, 3-13

information messages, 2-20

information operations (IO) capabilities, 2-32 through 2-38, 3-13

information operations (IO) objectives, 2-17, 3-3, 3-10 through 3-13, 4-1 and 4-2, 4-5, 4-8, A-4, A-23, A-25

information operations working group (IOWG), 4-6 and 4-7, A-1 through A-3, A-11

information superiority, vi, 1-1 and 1-2, 1-4 and 1-5, 2-1, 2-5, 2-16, 3-1 and 3-2, 3-7 through 3-10, 3-12 through 3-14, 3-16, 4-4 and 4-5, 5-9, A-5, A-7, A-9, A-22 through A-25,

intelligence preparation of the operational environment (IPOE), 3-2, 3-6, 3-8, 5-1 through 5-3, 5-5, 5-7, 5-9, A-6, A-11, A-15, A-18

intelligence support to IO, 5-3

IO annex, 3-16, A-6, A-24

IO concept of support sketch, 3-14, 3-17

IO intelligence summary, 4-8

IO planner, 1-5, 2-16, 3-1, 3-7, 3-10, 3-19, 4-6, 5-9

IO situation report, 4-8

J

joint special operations task force (JSOTF), 1-5, 2-3, 2-7, 2-12, 2-16, 2-18, 2-28, 2-30, 3-6, 3-19, 4-6, 4-8

K

key-leader engagements, 1-5, 2-1, 2-20, 2-23 and 2-24, 2-30, 4-4, 4-7, 5-9

Index

M

measure of effectiveness (MOE), 3-11, 4-2 and 4-3, 4-6
measure of performance (MOP), 4-5
message development, 2-22
military deception (MILDEC), vi, 1-2, 1-5, 2-1 and 2-2, 2-5 through 2-7, 2-9 through 2-11, 2-14, 2-19, 2-32 through 2-38, 3-1, 3-5, 3-13 and 3-14, 3-17 and 3-18, 4-7, A-1, B-1, C-3
Military Information Support operations (MISO), vi, 1-2, 1-5, 2-1, 2-11 through 2-15, 2-18 through 2-21, 2-24, 2-28, 2-30 through 2-39, 3-1, 3-5 and 3-6, 3-13 and 3-14, 3-17 and 3-18, 4-2 through 4-4, 4-6, B-3, E-1 and E-2
MISO programs, 2-11
MISO series, 2-12 through 2-15
mission analysis, iv, 2-12, 3-7, A-11
mission, enemy, terrain and weather, troops and support available-time available, and civil considerations, 1-5

N

noncombatant evacuation operations, 2-13

O

open-source intelligence (OSINT), 2-3, 3-8 and 3-9, 4-3, 5-2, C-2, D-1 and D-2
operation plan, 2-4, 2-28, A-6, C-3
operational-level OPSEC, 2-2
operations security (OPSEC), vi, 1-2, 1-5, 2-1 through 2-6, 2-8, 2-10, 2-12, 2-19, 2-32 through 2-39, 3-6 and 3-7, 3-13 and 3-14, 3-17 and 3-18, 4-7, A-1, B-4, C-1 through C-3
orders production, 3-15, A-6

P

peacekeeping, 1-4
physical attack, 1-2
physical security, 1-2, 2-32, through 2-38
priority intelligence requirement (PIR), 3-7, 5-3, A-12, A-15
propaganda, 2-1, 2-13, 2-26 through 2-29, 2-31, 2-33, 2-39, 3-6, 3-12, 3-18, 4-7, A-15, D-1, D-3 and D-4
public affairs (PA), 1-2, 1-5, 2-1, 2-12, 2-14, 2-18 through 2-21, 2-27 and 2-28, 2-30 through 2-32, 2-39 and 2-40, 3-1, 3-6 and 3-7, 3-13 and 3-14, 3-18, 4-2, 4-4, 4-7, A-1, C-3, D-1
public affairs officer (PAO), 1-5, 2-14, 2-19 and 2-20, 3-1 and 3-2, 3-5, 4-3, 4-6, D-1 and D-2
public information messages, 2-20

R

rewards programs, 1-5

S

Special Forces operational detachment A (SFODA), 2-8, 2-12 and 2-13, 2-16, 2-20, 2-27, 2-32, 3-1 and 3-2, 3-19
Special Forces operational detachment B (SFODB), 2-13, 2-16
special operations task force (SOTF), 1-5, 2-7, 2-12 and 2-13, 2-16, 2-27, 2-30, 3-19, 4-6, 4-8
staff estimate for IO, 3-2, A-22

T

tactical-level OPSEC, 2-2
target audiences, 2-11 through 2-14, 2-20, 2-28 and 2-29, 5-7, B-4, E-1
targeting process, 2-12, 2-20, 5-8
tasks against the adversary, 3-12
tasks to defend friendly forces, 3-13
tasks to shape the environment, 3-13
theme, iv, 2-12, 2-14, 2-21, 2-32, 2-36, D-3

W

working group, 2-3, 2-5, 2-28, 4-2 and 4-3, 4-6, A-1 through A-3, A-5, C-2
working with translators, 2-26

TC 18-06
22 March 2013

By Order of the Secretary of the Army:

RAYMOND T. ODIERNO
General, United States Army
Chief of Staff

Official:

JOYCE E. MORROW
Administrative Assistant to the
Secretary of the Army
1307001

DISTRIBUTION:
Active Army, Army National Guard, and United States Army Reserve: Not to be distributed; electronic media only.

PIN: 102{

www.ingramcontent.com/pod-product-compliance
Lightning Source LLC
Chambersburg PA
CBHW071209240526
45470CB00018B/1653